# Firefighters in Art and Media

Brent Allen Thale (photographer), Fireman statue by John G. Segesman in front of the Texas State Capitol building, Austin, Texas. Photo 2002
© Brent A. Thale. (For commentary, see entry no. 83)

# Firefighters in Art and Media

## A pictorial history

John A. Walker

Francis
Boutle
Publishers

London: Francis Boutle Publishers, 2005
First published by Francis Boutle Publishers,
272 Alexandra Park Road,
London N22 7BG
Tel +44 (0)20 8889 7744
info@francisboutle.co.uk
www.francisboutle.co.uk
in association with the Institute of Artology.
This edition is limited to 1000 copies.
Index compiled by Auriol Griffith-Jones.



British Library Cataloguing in Publication Data: a catalogue record for this book is available from the British Library.

Designed by Kathryn Tattersall of
Tattersall Hammarling & Silk,
Berkeley House,
163 Tottenham Lane,
London N8 9BT.

Printed in Spain.

ISBN 1 903427 23 1

# Contents

# Acknowledgements

Special thanks are due to Clive Boutle for his faith in my project and Kathryn Tattersall for her design skills, and to all those who have answered my requests for information and images: Colin Adrian; Kevin Arbuthnot, Andrew Hanson and Brian Saville of the West Yorkshire Fire Service; Derek Manley; Gary Archer of Leamington Spa Public Library; Joanna Barr, Conservation Project Manager, Artlab Australia; Ben Crane of tradecards.com; Alicia Chmielowski of the Fire Service Library of South Australia; Bill Seager of the South Australian Maritime Museum; Rachel Davey of the State Library of South Australia; Alan Griffin of Leamington Spa; Elizabeth King, research assistant at the Royal Academy of Arts Library; John Heiser, Ranger/Historian of Gettysburg National Military Park; Dr. Peter F. Kramml of Magistrat Archiv und Statisches Amt, Salzburg; Elizabeth Farningham; Carol Mitchell of The Potteries Museum & Art Gallery, Stoke-on-Trent; Kati Va'mos, Assistant Local Studies Librarian, Devon Library and Information Services; David Lewis, Curator of the Aurora Regional Fire Museum; Kate Perry, archivist at Girton College Library, Cambridge; Sarah Foden, information manager of Cadbury Trebor Bassett, Birmingham; the staff of the CIGNA Museum & Art Collection, Philadelphia; the Historic New Orleans Collection, New Orleans; Dime Novels and Penny Dreadfuls Collection of Stanford University Library; Ryan Hyman, Curator of Collections at the Macculloch Hall Historical Museum, Morristown, New Jersey; Commandant Raymond Deroo; Christian Decolloredo, public relations officer of the Brigade de Sapeurs-Pompiers de Paris; Charley Furey of firemanspictureframe.com; Matthew Withey Assistant Curator of Sculpture, Leeds City Art Gallery; Berit Wulstrup Jensen of the Picture Archive of Copenhagen City Museum; Jens Peter Olesen of graphicalic.dk; Deborah Wilson of the Chartered Insurance Company, London; Professor Josephine Withers; Dr Marcy S. Sacks of Albion College, Michigan; Eric Robinson, Reference Assistant of the New York Historical Society Library; Pauline LaRocco, Marketing Coordinator, Jockey International, Inc., Kenosha, Wisconsin; Mike and Barbara McCumisky and Dave Russell of the Fire Services Museum of Victoria, Melbourne, Australia; Dan Dozier of ToyTent.com; Robert Rocha, Assistant Fire Chief, Kansas City Fire Department; Sherrie Kline Smith of the Kansas City Public Library; David D. Altrichter of the Historical Preservation Committee, Slatington, Pennsylvania; Dale J. Fiore of the Evergreen Cemetery Association; Tim Gardner of glasgowsculpture.com; J. Whittington, Librarian of the Art Gallery of South Australia; Kevin Shea, retired New York firefighter; Susan F. Walker of the Oklahoma State University Library; Craig Eli Stone of the Art Asia Gallery, Milwaukee; Emma Strouts of Christie's Images Ltd., Jenni Lankford, Council Chamber docent, City Hall Gallery, Charleston, South Carolina; Michael P. Martin of the Onondaga Historical Association, Syracuse; Colin Vinnard, manager of International Services, Lions Gate Entertainment Inc.; Janice Braun, Special Collections Librarian, F.W. Olin Library, Mills College, Oakland, California; Wendy Morrow, Pictures Reference Librarian, National Library of Australia;

Matthews International Corporation, Pittsburgh; Kevin Corcoran of the Charleston County Public Library; Jerry Ripberger; Corbin Bronze Ltd.; Bill Jirousek of The Sculpture Center, Cleveland; Anne Woodhouse and Ellen Thomasson of the Missouri Historical Collection; staff of the Museum & Library of Maryland History, the Maryland Historical Society, Baltimore; The Richardson Library of the Saint Louis Art Museum; Susan Wilczak of the Krasl Art Center; *News Shopper* of Petts Wood, Orpington; Edward F. Flynn Jr., historian of the New Haven Fire Department, Connecticut; sculptor Hai Ying Wu; J. Gayle Camarda of the St Louis Public Library;

Sandy Barrie of Sydney; Suzy Gregson, Administrative Assistant, of the London Fire & Emergency Planning Authority; Eugene and Jeff Fairbanks; Mike Smith and Richard Crangle of the Magic Lantern Society; Henc R.A. der Roo, magic lanternslides collector; the local history department of Bradford Public Libraries; Robert Morphet of Wright & Sons, Bradford; Miriam Rogers Fowler, Curator of Education, Birmingham Museum of Art, Alabama; Richard Stansfield of York Castle Museum; Daryl Lang of *The Sentinel*, Carlisle, PA; Judith Caul of Guardian Syndication; Tina King, Director of Parks & Recreation, Harrisburg, Pennsylvania; artist John Dugger of San Francisco; Anita Gamill, Public Relations, STIHL Incorporated; Robert Workman and the agency Guy Chapman Associates; Shirley J. Packer of the historical archive of Imperial Tobacco Ltd, Bristol; Rhodri Morgan of the National Library of Wales, Aberystwyth; Julie Thies and Lisa Parrott Rolfe of the Speed Art Museum, Louisville, Kentucky; Marc Baron Tantillo (grandchild of Josef Josephu), of New York; Eva Treen of Jordan Memorials, Australia; Vivian Tillett of Wakefield Press, Adelaide; Patricia A. Eby-Manescu of the Slatington Public Library; Heinrich Krenn, curator of the Wiener Feuerwehrmuseum; Peter Y. Sobolev of St Petersburg; The Austin History Center, Austin, Texas; Katharine Oakes of the British Film Institute, London; Steve Taylor of Snaptastic. co.uk; Michael Kernan, Honorary Historian and Archivist of the Fire Service College, Moreton in Marsh, Gloucestershire; Stephen Voss, photographer, and FEMA (Federal Emergency Management Agency); Dennis W. Felty of Northstar Gallery; Mike Jorgensen, photographer; Brent Allen Thale, photographer; Ed Harvey of Fireservicephotography.com; Nikki Light, Rights Executive of Frederick Warne & Co./Ladybird Books Ltd; Peter Marshall and Matt Morgan of the United Firefighters Union of Australia; Sherrie Smith of Kansas City Libraries; Richard Margolis for his photograph of the sculpture by Hebard; sculptor Robert J. Eccleston for photographs of his Albany memorial; Regina Kaltenbrunner of the Salzburger Barockmuseum; Norman Taylor of the University of Washington; Martin Klein for his photographs of a Berlin fountain and memorial, and Stéphane Bauer and Stefan Sträubig for information about and images of the same sculpture; the painter Sergio Ostroverhy for information and an image; Edith Lahellec and Gilles Grandjean of the Musée Crozatier; Charles I. Manetta of Miami; Julie Anne Lambert of the Bodleian Library, Oxford; George Danhires, American sculptor of Ohio, for supply of slides; Evelyn Dow, Editor of the *Westside News*, New York; Public Monument and Sculpture Association, National Recording Project (UK) and the Save Outdoor Sculpture organisation, and the Art Inventories Catalog of the Smithsonian Institution Research Information System (SIRIS); Kathleen Herzog of www.sagbottom.com for supply of photos.

The wide-ranging research this book required would not have been possible without the existence of the Internet/World Wide Web and the Google search engine; consequently, I am indebted to their inventors and developers.

While very effort has been made to trace and acknowledge the copyright holders of the illustrations, the author and publisher will be glad to learn about any errors and omissions so that corrections can be made in any future edition.

# Introduction

Before the terrorist attacks on New York and Washington on 11 September 2001 (henceforth 9/11), heroism was unfashionable and Western culture was dominated by the worship of celebrities many of whom were selfish and shallow. Afterwards, there was a return to popularity of 'local' or 'everyday' heroes, that is, courageous public service workers willing to risk their health and lives to save others. New York's firefighters – who lost 343 members in the destruction of the Twin Towers – were one group of emergency workers who were praised and later honoured in art works, comic books, plays, television documentaries and exhibitions of portrait photographs. What many of their admirers may not realise is that there is a long history of representations of firefighters both real and fictional, although it was only during the nineteenth century that their role as heroes captured the popular imagination. This book documents that history through a chronological sequence of selected illustrations.

Organised firefighting can be traced back to Assyria in 850 BC and to Alexandria and ancient Rome. In the age of Augustus, corps of so-called '*Vigiles*' or night-watchmen were formed from both freedmen and slaves, who raised the alarm and extinguished fires. Seven battalions of 1000 men protected a population of one million. How the *Vigiles* dressed is not known because, unfortunately, a full-length portrait carving of one on a gravestone is indistinct due to decay. Probably, their uniform was a variation of that worn by Roman soldiers. More is known about the tools and equipment they used. The Roman Empire also gave rise to the patron saint of firefighters: St Florian, named after Florian of Lorch, a Roman army officer born in Cetium (now Austria) around 250 AD who rose to the rank of *Perfectus vigilum* or fire chief. Legend has it that Florian once prevented a town from burning with a single bucket of water. Florian converted to Christianity and when he refused an order to execute Christians he was tortured, set on fire and then drowned. There are several statues depicting him, and, naturally, he has been popular with Austrian artists. For instance, in Alter Markt (Old Market), Salzburg, Austria, there is a *Fireman's Monument* (aka *Floriani-Brunnen* or *Fountain of St Florian*) dating from 1734 consisting of a baroque-style fountain with a central column surmounted by a statue of St Florian sculpted by Joseph Anton Pfaffinger (1684–1758). During the late 1920s, Josef Josephu (1889–1970), a successful, academic-style, Austrian sculptor, who later moved to the United States, created a bronze statue of a Roman soldier – presumably Florian – emptying a bucket. Another bronze by Josephu represented a Roman soldier using a squirt to suck up water from a pail. Later, in 1935, Josephu created a larger than life-size bronze statue of St Florian wearing a helmet, body armour and holding a lance. This figure adorned the exterior of the Vienna Fire Department building until the latter was bombed during World War II. Fortunately, the statue survived and is now preserved in Vienna's Fire Brigade Museum, which occupies the old fire station in Am Hof, a large square.

Unknown photographer, Josef Josephu standing next to his bronze statue of St Florian outside the Vienna Fire Department, (1936). Photograph courtesy of Marc Baron Tantillo, New York

Josephu, who incidentally had an elder brother and fellow sculptor named Florian, created other bronze sculptures of contemporary Austrian firefighters, which are now in the Museum's collection. A relief image of St Florian also adorns a medallion produced by the San Diego City Paramedics. Stores selling Catholic religious merchandise also stock prayer cards with coloured images, resin figurines and medals featuring St Florian. One pewter statue on sale depicts a fireman with the taller figure of St Florian standing behind him like a guardian angel.

The literature on firefighting is huge and there are many archives and memoirs. There are also a number of published histories of the subject, some of which concentrate on famous fires, or firemarks, or fire appliances, or fire alarm boxes, or fire boats, or firefighters' uniforms, so no attempt will be made to repeat the whole story of firefighting. Such books are generally illustrated but their authors normally pay little attention to the iconography and medium of the representations or the artists who created them. (There are, in fact, few writings specifically about visual representations of firefighters; consequently, an iconographic study is overdue.) (1) Since I am an art historian rather than an ex-firefighter, my focus will be on those matters. Naturally, my text does include some information about the history and techniques of firefighting for which I am indebted to the aforementioned texts. (Of course, the images themselves are historic documents, which embody much information about firefighting in visual form.) Readers who wish to learn more about the history and methods of firefighting itself should consult the books listed in the Bibliography. What these illustrated histories also generally ignore are the numerous visual representations of fictional firefighters in books, magazines, cartoons and on stage and screen, which this book includes because they contribute to the public's perception of the occupation.

Research soon revealed the extraordinary variety of materials, media and forms – which encompassed the whole spectrum of culture, high, middle and low – in which representations of firefighters appear. For instance, there are balloons (a huge one depicting Harold the Fireman floats above New York's streets on Macy's Thanksgiving Day Parade), drawings, cartoons, caricatures, calendars, comic strips and books, glove puppets, fridge magnets, illustrations (in books, magazines and newspapers), paintings, murals, panoramas, dioramas, cycloramas, prints, portraits, sculptures, trophies, stained glass windows, memorials/monuments, badges, medallions, casino tokens, Steven graphs, magic lanternslides, photographs (of which there are tens of thousands), stereographs, cigarette cards, postcards, playing cards, phone cards, pin-ups, posters, postage stamps, decals and stickers, T-shirts, tattoos, wallpaper, films (documentary, fictional and animated), television documentaries and drama series (Richard C. Yokley's book on the latter subject discusses over 75 examples), advertisements, trade cards, toys, pocket watches, penis pumps, musical water globes, porcelain and bronze figurines, tableware (there are drinking jugs, salt and pepper pots and teapots in the shape of firemen), weathervanes, thimbles, novels, poems, musicals and plays. Last but not least, there are video games; for example, *Rosco McQueen, Firefighter Extreme*, Sony Playstation, published by Psygnosis in 1998.

Hundreds of examples of so-called 'firematic antiques' are illustrated in James and Sandra Piatti's book *Firefighting Antiques & Memorabilia* (2001), in Andrew G. Gurka's *Firefighting Collectibles* (2001) and in Mihály Kútvölgyi's *Hungarian Fireguard Mementoes* (2001). The number of items utilising firefighter imagery currently on sale is extraordinary: the Yahoo! shopping website lists over 400 suppliers selling over 1000 different products and the Ebay auction websites lists thousands of firefighter-related items for sale.

One cannot even describe firefighters as 'unsung heroes' because there are numerous songs about them. To cite one recent example: Rockabilly legend Carl Perkins' 'The Fireman', which was written with Bob Johnston and recorded just before Perkins died in 1998 but only released by JAM Records after 9/11. In the nineteenth century, the lyrics and music for marches and two-steps were issued via music sheets with covers adorned with firefighting scenes.

There is also a minor but thriving artistic genre known as 'firefighter art' or 'fire art'. Many contemporary artists – both men and women, both professional and amateur – depict firefighters in an illustrational manner. Their primary audience appears to be the firefighting community. Others produce paintings and prints to assist charities for firefighters. Certain artists, like Allan Albaitas of Las Vegas and Paul Combs of the

Josef Josephu, *Austrian Firemen*, (late 1920s). Sculpture, bronze. Vienna Fire Brigade Museum. Photo courtesy of Heinrich Krenn

Kent Meireis (photographer), *New York, New York: Harold the Fireman gets lined up for the 75th Macy's Thanksgiving Day Parade*, (22 November 2001). © Kent Meireis/The Image Works, © 2004 TopFoto.co.uk

Bryan Fire Department, are veteran firefighters or, like the Canadian Captain Paul Beckingham and the Parisian painter Sergio Ostroverhy, retired firefighters. In some cases, artists generate prints from their drawings and paintings to market via the World Wide Web. For instance, Lisa Chadwick of Florida, who makes drawings with graphite, and Jean Miller Erskine a self-taught American painter. Since 9/11, a favourite composition of American artists has been the patriotic one of firefighters juxtaposed against the Stars and Stripes.

In selecting illustrations for this book, I have tried to reflect the variety of media and subjects found in representations of firefighters. Most illustrations are American and British because of the sheer number that exist, the wealth of information about them and the long tradition of firefighting in both nations. Anyone seeking images from Australia, France, Germany and New Zealand can consult Colin Adrian's *Fighting Fire! A Century of Service 1884–1984* (1984), Raymond Deroo's *La Fabuleuse Histoire des Pompiers* (2002), Jens-Peter Wilke's *150 Jahre Berliner Feuerwehr, 1851–2001* (2001) and Gavin McLean's *New Zealand Tragedies: Fires & Firefighting* (1992).

Visual representations of firefighters are addressed to different groups and ages. For example, much material is produced for infants, children and teenagers: it includes illustrated reading and colouring books, 'show-and-tell' books (used as visual aids by teachers), 'action figures' and model fire engines to play with, and costumes in which to dress up and act out roles.

Many representations are directed towards firefighters and their families. There are in fact so many fire departments around the world that firefighters constitute a major subculture and, naturally, I hope this book will be of interest and value to them. The functions that representations fulfil for firefighters, their families, friends and local communities are rather different than their function for the general public and regular gallery goers. For instance, a sculptural monument dedicated to fallen firefighters where annual ceremonies of remembrance are held following an initial dedication event is not primarily a site of aesthetic experience. It is, rather, a site of emotion, grieving, memory, catharsis, renewal and solidarity with a particular community; consequently, such sculptural monuments cannot be judged by their artistic quality alone, which to be frank is sometimes mediocre or stereotypical. As we shall see, most such monuments are traditional in style – they continue the naturalistic style associated with war memorials and statues of generals and politicians made by academic artists during the nineteenth century, which in turn derived from the realist portrait statuary of ancient Rome and the Italian Renaissance. This is why firefighter memorials – most of which are funded by firefighters and local communities – are generally ignored by the contemporary art world, that is, by art magazines, books and galleries devoted to modern and avant-garde art. (The artists who created the memorials reciprocated by ignoring modern art. Virtually the only avant-garde works featuring firefighters I have discovered are Julio González's 1933 welded metal abstracted sculpture of a fireman called *Pompier* and Red Grooms' 1959 mixed-media happening entitled *The Burning Building* and Paula Rego's 1966 painting *The Firemen of Alijo*.) Surely, this is an unfair and, art-historically speaking, an untruthful state of affairs. This book includes instances of living artists, some self-taught, who are highly successful in terms of public commissions and yet receive no evaluation or recognition from the art world. Information about them can often be found on their own websites or the websites of local fire departments.

Another non-artistic function which representations of firefighters serve is fund raising for charitable purposes. For example, virtually every maker or publisher of images associated with 9/11 declared that all or a proportion of the sale price would be donated to firefighting charities intended to help injured survivors and the families of the fallen.

When selecting reproductions, I excluded memorials without firefighter figures; for instance, the American National Fallen Fighters Memorial at Emmitsburg, Maryland, dedicated by Ronald Reagan in 1981, consists of a stone cairn surmounted by a Maltese Cross surrounded by name plaques. The Maltese Cross, which has eight points and four equal arms that expand in width outward, is a symbol popular with firefighters (and First Aid organisations) because it has long been associated with saving lives and extinguishing fires. In the Holy Land, Saracens battling Crusaders used firebombs against the Knights of St John – who later lived on Malta – and so some had to become expert in fighting fire and in rescuing and aiding victims. Today, the Maltese Cross appears on badges, flags, lapel pins, medals, memorials, patches, jewellery and wall plaques, and some items have relief decorations of firefighting paraphernalia. In the United States, frames for photographs have also been manufactured in the shape of the Maltese Cross. Head-and-shoulder portraits of New York firemen were inserted into a circular frame at the centre of the Cross and the whole object rested on a small stand. In the example reproduced, the fireman portrayed is wearing a Maltese Cross badge on his dress cap. Maltese Crosses are frequently worn by firefighters as badges of honour and as lucky charms, and they have been adopted as logos by numerous American fire departments.

Another striking memorial without sculptures of firefighters was erected in 1976 inside the Cathedral Church of St John the Divine, New York. It is made from a long beam of charred wood held aloft horizontally by a skeletal hand of steel. A second vertical beam completes the shape of the Christian cross. The structure rests on a stone cairn, which has a plaque inset. The *Memorial to the Fallen Twelve* (who died during a Broadway blaze in October 1966) was conceived and constructed from materials found at fire scenes by Ralph Feldman, a retired firefighter and Fire Marshall.

I have also excluded images, often called 'inspirational' or 'motivational', which are excessively sentimental and maudlin, that are, in fact, quasi-religious kitsch. For example, in the wake of 9/11, when emotions ran very high, Greg and Tim Hildebrandt, two American brothers who are successful illustrators, produced paintings, which were later reproduced and sold as wall posters, with titles such as *The Day the Angels Cried*. They showed exhausted and grieving New York firefighters being consoled by young female angels glowing with Heavenly light. In addition, Hollywood-style movies such as *Firestorm* (1998), *Ablaze* (2000) and *Trapped* (2001) have been omitted because they are so poor.

I admire firefighters because I find it astonishing so many young men and women are willing to enter a profession with such a high risk of injury and death. The memorials listing the names of the fallen inside and outside fire stations are a constant reminder to the living of those who have made the ultimate sacrifice. Politically, I also support public sector workers who are generally poorly or modestly remunerated for performing tasks that are essential to society. Because of low wages, some firefighters are unable to afford the house prices in expensive areas and so have to travel long distances to work. Others have to take second jobs to survive. Many

Molly Higgins (photographer of photo frame), *Fireman Photograph in Maltese Cross frame*, (n.d.). 23 x 23 cms. Source: J.G. & S.F. Piatti, *Firefighting Antiques & Memorabilia* (2001)

firefighters belong to trade unions, some of which have banners, and although strikes for higher pay are rare, as we shall see, they have occurred and been depicted in cartoons and photographs.

At the same time, images of firefighters need to be examined critically because some have been racist and many are clichéd and stereotypical. Analysis of a large corpus of representations soon reveals repetitions such as the fire chief portrait, the group portrait, the rescue or saved motif, the hose shot, the departure from the fire station, the speeding fire engine, the sliding pole motif, the firefighter's lift or carry, the firefighter on a ladder, and so forth. Furthermore, the subject raises issues of gender and race. (White male firefighters have been guilty of racism and misogyny.) It is for this reason that images of firewomen and fire officers from ethnic minorities have been included.

Naturally, I hope this book will appeal to artists, art students and art history lecturers and students, and to the millions who demonstrate an interest in the visual arts by visiting museums and galleries. Such readers will encounter little-known works of art by famous artists, and they will learn about unfamiliar art works and artists. The images and commentaries will act as windows on the subject of firefighting and the portraits will introduce them to some of its principal historic characters. Part of my intention in writing this book was to contribute to labour history.

Because fire is such a useful yet dangerous and destructive natural force, it is of intense interest to humanity. The appeal and excitement of out-of-control fires to the popular imagination during the modern era is obvious. One thinks of the visual spectacle of major conflagrations: flames, billowing smoke, collapsing buildings, desperate victims, red fire appliances or trucks racing through the streets with their bells ringing or sirens wailing, the turntable ladders thrusting high into the sky, the long hoses issuing powerful jets of water. The phallic/urination associations of water hoses and nozzles are obvious and these are sometimes made explicit in actual depictions. For instance, one American sticker shows a naughty boy wearing a fire helmet peeing on a fire. On his helmet are the words 'Fire Dept', plus the safety advice 'Put 'em out'. Another example is an image on a British postcard showing a crowd admiring a powerful stream of water from a hosepipe that seems to be emanating from a fireman's groin. One spectator remarks admiringly: 'Blimey – what a man!' The clothes and equipment favoured by firefighters are also attractive – indeed have a fetishistic appeal. Many people find figures dressed in uniforms sexually stimulating.

Unknown artist, *Blimey – what a man!* (n.d.). Colour picture postcard, 9 x 14 cms. Holmfirth, Yorkshire: Bamforth & Co. Ltd., Comic series no. 868. Author's collection

Because firefighters perform much of their work in public, it is highly visible profession and this explains some of its popular appeal. Since the advent of film, television and satellite communication between continents, millions who are far from the scene of a major disaster can also watch rescue workers in action. In addition, fire brigades have organised events open to the public such as balls, street parades, demonstrations, exhibitions, reviews and tournaments. In Britain, some of these took place in the presence of royalty. In 1903, for example, an International Fire Exhibition was held at Earl's Court, London. On show, besides firefighting equipment, were paintings, drawings, sculptures and prints depicting fires, firemen and rescues. Simultaneously, at the Empress Theatre, daily performances were given of a popular entertainment entitled 'Fighting the Flames'.

Important European artists such as J.M.W. Turner, Wenceslaus Hollar and Philippe de Louthenbourg have generated distant views of significant fires, but in such representations, the focus is generally upon the burning buildings or cities rather than upon people. The latter are usually depicted as fleeing crowds or as spellbound spectators (fires have given rise to a category of regular watchers known as 'fire buffs' some of whom have formed collections of firefighting antiques). Many paint-

Joseph Hoffman Johnson, *New York Firemen rescuing Samuel J. Tindale*, (c. 1850). Oil painting on a fire engine's wooden side panel, 41 x 65 cms. Hudson, NY: American Museum of Firefighting. Photo courtesy of Museum

ings and prints of fires do not show any firefighters at all. Others depict them in teams at the bottom of the composition where they tackle the blaze by various means. Most of the images chosen for this book concentrate on the firefighters themselves rather than upon fires.

Throughout history, the majority of firefighters have been young fit men attracted by the adventure, glamour, risk, camaraderie and job security of the profession. In fact, it has been a calling or vocation as much as a profession. As in the case of ancient Greek and Roman mythological figures and personifications, nineteenth-century firefighters were associated with a number of objects or attributes: an axe, a hook, a helmet or parade hat, a uniform plus a cape, an arm badge, a brass torch, a brass hose nozzle, belt buckles, braces or suspenders, a speaking trumpet and spanner wrenches (to couple and uncouple lengths of hose). Many such items are now antiques and are preserved in the numerous fire museums that exist in developed countries. Those in private hands are valuable collectibles.

As the profession developed, the equipment used became more elaborate, expensive and complex. Great pride was taken in the water pumps that were manufactured during the eighteenth and nineteenth centuries. Some of them were even decorated with elaborate, baroque-style ornamentation and painted vignettes. In New York, the English-American artist Thomas Sully (1783–1872) was well paid for painting allegorical or patriotic scenes, and portraits on the side panels of wooden fire engines. Other American 'fire engine' artists were: L. Ayer, M. Betsch, John Carolan, David Rent Etter, Henry Inman, William H. Philip, T. Pine, Charles Peale Polk, John Vanderlyn and John Archibald Woodside. Paradoxically, panel paintings depicting firefighting scenes were in the minority. One American example, painted by Joseph Hoffman Johnson (1821–90), recorded the desperate rescue effort that followed a huge boiler explosion in a machine shop in Hague Street, New York on 4 February 1850 in which 67 died. Samuel J. Tindale, the 15–year-old son of a prominent citizen, was one of many trapped in the smouldering ruins. After 20 hours, firemen from several different companies succeeded in extricating him but Samuel died soon afterwards.

In addition, artists decorated buckets, oilcloth capes and parade hats (stovepipe hats). Michael J. Novak reports that African-American artists such as David Bustill Bowser (1820–1900) and Robert Douglas painted lettering and scenes on parade hats. (2)

Thus, in some instances, artists were closely associated with the firefighting profession. Not only did firefighters act as patrons by paying hundreds of dollars to painters for panel pictures and portraits, certain artists were themselves volunteer firemen. For instance, the American artist Julian Scott (1846–1901), who in 1881 painted an elegant portrait of a fireman – probably a foreman – holding an ornate speaking trumpet posed against the wheels of a fire engine, was a member of Plainfield Fire Department, New Jersey. Moreover, as we shall discover, one seventeenth-century Dutch artist – Jan van der Heyden – was even a noted inventor of firefighting equipment and the author of an early manual on firefighting methods. James Colquhoun, a seventeenth-century Glasgow mason, builder and sculptor, has also been credited with inventing the City's first fire engine. Ando Hiroshige (1797–1858), the famous Japanese printmaker whose work inspired Vincent van Gogh, was the son of a fire warden and spent time in his youth working as a fireman in Edo.

A whole book could be devoted to the invention and evolution of firefighters' headgear: caps, hats and helmets. The latter were designed to protect them from falling debris and water. American and British ones during the nineteenth century differed considerably. The former had extended rims at the back to protect the neck and most had front shields or nameplates for identification purposes. Again, some American fronts were decorated with hand-painted images of fire engines and portraits of firemen.

The archetypal British helmet made from brass or white or 'German' metal introduced by London's Fire Chief Captain Shaw in the 1860s, which remained in use until the 1930s, was based on ornate crested ones that he had seen military firemen wearing in Paris. In France, firefighters are called '*sapeurs-pompiers*' (sappers or engineers and fire or pump men). Napoleon established the first corps of *sapeurs-pompiers* in 1811. A whole category of art displayed at the annual Salon exhibitions in Paris devoted to classical subjects was sarcastically dubbed 'pompier' because there were so many paintings with figures sporting firefighter-type helmets. More recently invented types of equipment are less useful to artists; for instance, breathing apparatuses have masks that obscure the faces of firefighters. Of course, if an artist wishes to avoid identifying individuals and their gender and race for reasons of generality, then facemasks are helpful in this respect.

The representations in this book could have been arranged according to the medium employed: painting, photography, sculpture and so on. However, while it is clear there is a historical sequence of media in which new ones such as film and television emerge to challenge or even reduce the power of older ones, today many media of different ages coexist. For instance, stone and bronze monuments, which originated in ancient times, are still being produced even in the age of digital imagery. As we shall see, many such monuments were created out of doors across the United States during the second half of the nineteenth century and in the period 1990–2003. Recent examples usually take the form of bronze figures executed in a naturalistic style. Virtually every American state and major city now has one or is planning one. Europe, in contrast, has far fewer examples.

Furthermore, although each medium possesses certain unique characteristics, in other respects, they overlap: portraits, for example, are to be found in paintings, photographs, prints and sculptures. This often means that an iconic image first developed by a painter is later copied or duplicated by a photographer, printmaker or sculptor. In this way, continuities occur between media and over time. Only by comparing many examples from different media, can such common elements be identified and traced.

The vast majority of firefighter representations capture dramatic moments and present firefighters in a positive light. One should perhaps point out that firefighters have to confront the charred bodies of victims and colleagues they were unable to save, but few images of grisly remains or wounds caused by burns are made public. In addition, many of the tasks performed by firefighters do not involve putting out fires: answering false alarms, dealing with hazardous materials, educating the public and schoolchildren about safety issues, giving medical attention, rescuing people and animals from floods or cutting the injured – using the so-called 'Jaws of Life' or Hurst tools – from crashed cars and railway carriages, for example. The development of firefighting techniques and equipment over time gave rise to specialisation such as engine, ladder and Haz-Mat companies; some modern fire companies do not put out blazes because they are rescue-only. Furthermore, fighting oil well, bush or forest fires calls for different methods than those used in towns and cities. Again, I have tried to reflect the diversity of tasks firefighters perform in my choice of illustrations.

In addition, firefighters spend many hours inside firehouses or stations waiting for alarm calls. (This has given rise to a negative view of firefighters as a lazy group of workers.) To occupy themselves, firefighters clean and maintain their equipment, exercise to keep fit, train and drill to hone their skills, cook meals, study and engage in hobbies. A few photographers have documented the routine life inside American firehouses; for example, in 1943 the African-American Gordon Parks did so. In 1977, another American photographer – Jill Freedman – published a book of photographs of firefighters attached to a New York firehouse. A more recent study, entitled *Firehouse* (2002), written by historian and journalist David Halberstam, provided an intimate portrait of the men and families of Engine 40, Ladder 10, two New York companies that lost over ten firefighters on 9/11. As we shall see, the daily life and interpersonal relationships between fire personnel inside stations has formed the basis for several television dramas in Australia, Britain and the United States.

Regrettably, not everyone in society regards firefighters as heroes performing a vital public service. For instance, in divided communities such as those of Northern Ireland, firefighters and paramedics have been attacked because one faction was determined to prevent assistance reaching their injured opponents. The fact that firefighters wear uniforms like the army and police, and are part of local government, means that some dissident groups and rioters perceive them as authority figures maintaining the social order and so they act in a hostile manner towards them. Sometimes, during civil disturbances, firefighters are ordered to control members of the public rather than to assist them. As we shall discover, in 1963, Alabama firefighters were instructed to turn their hoses on civil liberties demonstrators.

A few firefighters also fail to live up to the high standards of their profession. For instance, in 1991 Captain John Orr, an experienced firefighter and fire investigator for the Glendale Fire Department in California, was sentenced to life imprisonment after being convicted of several acts of arson that caused fatalities and massive loss of property. In 2002, Joseph Wambaugh wrote a book about him entitled *The Fire Lover: A True Story* and, in the same year, the telemovie *Point of Origin*, starring Ray Liotta, was screened. Such a rogue firefighter also appears in the 1991 feature film *Backdraft*. Another negative movie portrayal of a firefighter can be found in the 1997, made-for-Cable Television film *Bad Day on the Block* (aka *Under Pressure*) starring Charlie Sheen. He plays a fireman-hero decorated for saving a baby from a blaze but whose mental health is deteriorating. Soon he terrorises his neighbours and becomes a psycho-killer. In the past, fireman have also been depicted as violent or as drunkards or as lechers. Furthermore, once they withdraw their labour and stand on picket lines, they soon lose their heroic status in the eyes of conservative newspapers and government ministers (even those belonging to a so-called labour party) unwilling to authorise wage increases. As one British fire chief put it at the time of a strike during 2002–03: '... we were getting nothing but praise. Now they are demonising us. We've gone from heroes to zeros.' (3)

In the main, however, and especially since the sacrifices of 9/11, the vast majority of the populace regards firefighters as heroes. (It should be said that most firefighters are modest and feel embarrassed by public praise. When hailed as heroes, they tend to respond by saying 'we were only doing our job'. Perhaps one can say the profession is itself inherently heroic.) Their own tradition of duty and sacrifice demands that they live up to these ideals and, as the images from the nineteenth century illustrated in this book show, a script was written and roles established for them long ago.

One wonders if the fresh respect in which public sector workers are held will last because, at the time of writing, celebrity culture seems to be reasserting itself. What remains certain is that anyone trapped in a burning building or a crashed vehicle will scream for firefighters not the latest overpaid vapid celebrity.

1 See Bibliography for writings by Chief Ronny J. Coleman, Robyn Cooper, Amy S. Greenberg, Kenny Hunter, Michael J. Novak, Viktoria Sargent and Richard C. Yokley.

2 Michael J. Novak, 'The American volunteer Fireman and the Daguerreotype,' *The Daguerreian Annual* 1997, (Pittsburgh, PA: The Daguerreian Society, 1998), pp. 88–137.

3 Dave Smith, quoted in, 'Eight days in the life of a strike,' by Blake Morrison, *Guardian G2*, January 21, 2003, pp. 1–5.

**A note on terminology**

The term 'firefighter' gained currency during the 1970s and 1980s for two reasons; first, the word 'fireman' included railway employees who stoked the fires heating the boilers of steam locomotives and firefighters wanted to make it clear that they fight fires not stoke them; second, women began to join the fire service and so 'fireman' was no longer accurate or appropriate. The domination of the fire service by men who remain boys at heart resulted in nineteenth-century American personnel being referred to affectionately as 'fire laddies' and this masculine emphasis continues into the present in such expressions as 'The Guys', 'The Boys' and 'The Brotherhood'. A gender-neutral term was 'smoke-eaters', which derives, of course, from the smoke firefighters had to inhale inside burning buildings before the advent of breathing equipment. (Australian firefighters refer to themselves as 'firies'.) In this book, I have tended to use the words 'fireman' and 'firemen' when writing about earlier centuries and the term 'firefighter' for more recent times. However, given the additional tasks and responsibilities the occupation has acquired over the years – such as medical ones – even the term 'firefighter' is itself now too limited a description.

Because giving one's life for another is the ultimate act of human love ... and because the human conscience is the ultimate tribunal of right and wrong, the firefighter's commitment to duty and sacrifice rests upon love and morality ... [which] have been the spiritual and ethical pillars of enlightened societies throughout the broad sweep of human history. So, the firefighter's commitment to duty and sacrifice carries a civilizing and ennobling message beyond the fire service to society at large. Donald Martin Reynolds, *Monuments and Masterpieces* (1997), pp. xv-xvi

# Firefighters in Art and Media

## Catalogue of images

In the entries that follow, no references/footnotes appear but the names of authors are mentioned so that readers can find the sources in the Bibliography.

1 Unknown artist, *Fire at Tiverton*, (1612?). Woodcut illustration, published in a tract entitled *Wofvll Newes from the West-parts of England*

## 1 1612

## Fire at Tiverton

The illustration reproduced here has been variously dated and has been used to picture a fire in Tiverton, Devon on Wednesday 5 August 1612. Images generated by the woodcut relief process first appeared in ancient China. European examples date from the thirteenth century. Cutting into and removing wood from a smooth block to create an image and then inking the remaining raised surfaces to generate paper prints is thus a venerable method of producing rather crude but charming reverse images to illustrate books and religious pictures.

In 1612, Tiverton was a centre for the wool trade in which 8,000 people were employed. Old chronicles report that the cause of the fire was a furnace in a wool dyer's workshop that had been neglected by an apprentice. It quickly spread and destroyed most of the town. The loss of 600 homes and businesses, goods and provisions was immense; consequently, the townsfolk were left impoverished.

The woodcut image, in fact, may date from earlier than 1612 and may well have been a generic fire scene that was used to illustrate many fires described in chapbooks (small popular books sold by chapmen or peddlers). The image is unrealistic in the sense that it does not convey the scale of the disaster at Tiverton. However, it does inform us that firefighting methods at the time were primitive: leather buckets (householders were expected to maintain a supply) were used to carry water up ladders and iron hooks were used to smash windows and pull down thatched or tiled roofs. Hooks were employed to vent the fire – to let heat and smoke escape – and to stop it spreading to neighbouring buildings. In this instance, there is one hook attached to three shafts so that three men could pull on it at once. The use of hooks and ladders probably dates back to military warfare involving sieges and, in the future, some fire companies came to be called 'Hook and Ladder Companies'.

Although the firemen are dressed alike, in this case they do not seem to be wearing any special uniform. They were probably all local volunteers. However, they do seem to have a foreman (centre) who has his arms raised like the conductor of an orchestra. Another figure of interest is a man who is carrying goods away from the blaze. For many years, such porters or salvage men were as important as firefighters. To modern eyes, it seems strange that the firemen are not attempting to rescue two individuals inside houses; they are left to fend for themselves and one is using a rope to lower a child to the ground. The fire has attracted a crowd headed by a priest who prays for divine assistance. (Ironically, the day of the fire had been appointed by King James I to be spent in prayer and other spiritual exercises.) Some Puritans considered this fire and others in Tiverton to be God's punishment for the inhabitants' failure to observe their religion with sufficient zeal.

## 2 — 1762

## William Hogarth, The Times, Plate 1

Hogarth (1697–1764) was an English painter, printmaker and moralist who became noted for his portraits, popular narrative print series and his crowded London street scenes. His engraving *The Times*, Plate 1, published on 7 September 1762, depicts a dramatic and chaotic urban scene in which fires are destroying buildings but householders located at windows on the left are using metal squirts (syringe-like devices) to direct jets of water at a fireman below who is trying to fight a blaze.

This image, in fact, is a political satire with a host of personifications and symbolic references to contemporary people, places and events rather than a naturalistic record of an observed event. Comprehending the details and characters of the satire is now rather difficult, especially since there are two versions of the print with different characters portrayed.

Briefly, it seems Hogarth was siding with King George III who was trying to extinguish the blaze of the Seven Year War in Europe, then being fanned by the Whig politician William Pitt 'the elder', who in this version of the print is disguised as Henry VIII (the man on stilts). The principal fireman standing on a pump directing water from a fixed 'goose-neck' nozzle towards a globe of the world is thought represent Prime Minister John Stuart, 3rd Earl of Bute, or the King. (Hogarth was to be savagely attacked by John Wilkes in the pages of the *North Briton* for this print.) The many other references to contemporary events are irrelevant to the theme of this book. (Those interested in the political allusions can consult art-historical texts on Hogarth.) From our per-

2 William Hogarth, *The Times*, Plate 1, (1762). Engraving. Photo courtesy of the British Museum, London

spective, the print is useful for the information it contains about firefighting in eighteenth-century London.

At the centre of the composition is a hand-drawn and hand-operated wooden pump – a cistern with a piston pump inside – with levers on either side of the main body of the appliance, which six or seven firefighters are pressing up and down. (This pump may have been one made by Elizabeth Nuttall of Long Acre because an eighteenth-century broadsheet advertising her appliances is headed by an image that shows a virtually identical view of a pump and firemen to that employed by Hogarth.) A suction pipe is drawing water from a source in the foreground but buckets are also filling the well of the pump. Two firemen, each carrying two buckets, scurry back and forth. On the side panel of the pump, there appears the emblem of the Union [Insurance] Office founded in 1714, that is, four clasped hands. Consequently, the firemen were probably members of one of the private brigades, which had been organised by insurance companies in London after the Great Fire of 1666. However, some of them appear to be soldiers and sailors, and some are dressed as Scottish highlanders (they were followers of Lord Bute).

The water battle in progress may also indirectly reflect the rivalry that used to occur between competing brigades employed by different insurance companies. The initial aim of these companies, it should be stressed, was to save insured properties (not property in general) rather than human life. It was for this reason that fire escape societies were formed in London during the early nineteenth century.

Hogarth adapted the subject of firefighting for his own allegorical purposes; nevertheless, his print tells us much about the methods of the fire service in his lifetime.

## 3 1690

## Jan van der Heyden, Fire in a Soapery

Jan van der Heyden or Heiden (1637–1712) was a Dutch artist who lived in Amsterdam from 1650 and became noted for painting still-lives, landscapes and townscapes (both factual and imaginary). Remarkably, he and his son Jan were also appointed Fire Master Generals of Amsterdam. He is reputed to have become wealthy by inventing the fire pressure hose (to replace hauling by buckets) and by designing, patenting and manufacturing improved versions of fire engines and street lighting. To make the 50–feet-long hoses, which were then linked together with metal couplings, strips of leather or sailcloth were sewn together. Their length and flexibility meant they could be taken up a ladder and inside a building to reach the source of the fire. For this reason, they were called 'worms' or 'water serpents/

3 Jan van der Heyden, *Fire in a Soapery*, (1690). Etching/book illustration

snakes'. William of Orange brought several Dutch engines with him when he became King of Britain and Ireland in 1689. With his son, van den Heyden wrote and illustrated with engravings an important early treatise on firefighting: *A Description of Fire Engines with Water Hoses and the Method of Fighting Fires now used in Amsterdam*, (1690; 2nd edn 1735; English translation 1996).

Some of van der Heyden's illustrations were dramatic depictions of conflagrations. These were intended to show the inadequacy of earlier fire engines and firefighting methods. Other illustrations were precise, detailed and rather dry. The latter resembled technical drawings more than works of art and can be viewed as advertisements for his products because their focus was upon the pumps, hoses and methods rather than upon the deeds of firefighters. Indeed, at this stage, the latter were not represented as heroes but as anonymous operatives. (The Dutch firefighters were organised into companies headed by fire chiefs and foremen located in local districts. They came from various guilds and construction trades and were appointed by the city; consequently, they were not volunteers.) Nevertheless, one sees in the way pumps – which at this stage did not have wheels (they were carried by hand or dragged on sleds, or placed on wagons) – are operated by teams of men

4 Unknown artist, *Phoenix Fireman*, (1780s). Paint on wooden bargeboard. Photo courtesy of Phoenix Assurance Company

pressing down on levers on either side in order to suck and discharge water through the hoses that heavy physical labour was involved. Men could only perform this task in short bursts and so, until the advent of steam-powered pumps, many men were needed.

In the engraving reproduced, van der Heyden depicts a fire that occurred in an Amsterdam soapery called the 'Bruninvis' on 18 April 1682. (The letters in the image refer to items in a plan of the buildings in the area reproduced below the main scene.) It seems the fire started in a boiler and then spread to attics filled with peat. Three peat carriers 'who had been assigned to a fire engine' ran to get it and the fire was rapidly brought under control. As the illustration demonstrates, the engine plus hoses enabled the firefighters to climb up on to the roof of the building and attack the fire from two directions. Van der Heyden's descriptions of fighting this and 59 other fires were designed to prove the effectiveness of his engines, methods and organisation to the Governors of Amsterdam. Some of his drawings for the etchings are preserved in the Amsterdam Historical Museum and others in the Print Room of the Rijksmuseum.

**4** **1780s**

**Unknown artist, Phoenix Fireman**

**5** **1800s**

**Unknown artist, Sun Fire Office advert**

This painting, executed on a shaped wooden support, promoted the Phoenix Fire Office, which was founded by sugar bakers in London in 1782. It has a vignette with an allegorical female figure in the centre holding a shield featuring the Company's miraculous firebird emblem. Behind the figure is a night-time firefighting scene, which includes firemen directing hoses and operating a hand pump plus, in the distance, the Monument in the City of London commemorating the Great Fire of 1666. Flanking the vignette are two firemen holding axes and dressed in red uniforms. At each end, there are views of the River Thames and this indicates that the painting was a 'backboard', which was mounted on a barge owned by Phoenix.

Similar symmetrical compositions can be found in the images that used to head certificates and policies issued by early insurance companies. For example, in the

5 Unknown artist, *Sun Fire Office advert*, (1800s). Photo Royal & SunAlliance PLC

case of the Sun Fire Office founded in London in 1710, one finds depicted the Office's sun firemark, a fireman plus his tools, a salvage man, and a primitive wooden pump. In the case of the Sun image (a sun with a face surrounded by stylised rays alternatively straight and wavy to signify light and heat), the blue uniform and leather helmet are very similar to that of the main figure in William H. Pyne's aquatint of 1805 (see below). The salvage man has a red coat and wears the Company emblem on a chain round his neck rather than on his sleeve. The sack on his back crammed with goods is also adorned with a huge Sun Company emblem. The fact that the two figures facing one another are of equal size implies that, at this time, salvage men were considered as important as firemen. Clearly, given the opportunities for pilfering, it was vital that such men were reliable and trustworthy.

Early images of firemen, therefore, were often employed in what were in effect corporate advertisements. During the twentieth century, the tradition persisted because many adverts issued by fire insurance companies have featured images of firefighters.

## 6 — 1792

## J.M.W. Turner, The Pantheon, Oxford Street, The Morning After the Fire

## 7 — 1800

## Unknown artist, Fireman of the Westminster Fire Office in Uniform

As mentioned in the introduction, Joseph Mallord William Turner (1775–1851), one of England's most famous, prolific and innovative artists, was fascinated by the spectacle of major fires. His interest in the subject dated back to his youth because several firemen appear in his 1792 drawing and watercolour *The Pantheon, Oxford Street, The Morning After the Fire*. The Pantheon was a large suite of assembly rooms or 'indoor pleasure resort', situated between Oxford Street and Great Marlborough Street, London. A lavishly decorated theatre and a ballroom were included. It was designed by the architect James Wyatt in a neo-classical style and built between 1769 and 1771. On 14 January 1792, the Pantheon caught fire and was reduced to a shell. Local residents were alerted to the danger by the noise of wooden rattles or clackers used by nightwatchmen for the purpose. Clearly, the fire brigade that arrived on the scene had not been able to stifle the blaze at an early stage.

Turner, who was then a young man of 17, does not depict the fire itself but its aftermath, in particular the ruined architecture. In front of the building's façade and portico, curious onlookers from various social ranks have gathered to peruse the ruins and to gossip. In the middle distance firemen gather up hoses while in the bottom left-hand corner one firemen kneels to fill buckets with water from a stand pipe or fire plug while a second standing fireman uses them to clean the wheels of a wooden engine. Nearby, a well-dressed man observes. The engine, which is covered with icicles because of the cold, is surely one of the manual types manufactured by John Gray or Richard Newsham during the eighteenth century. The standing fireman wears a blue uniform which has an arm badge and he is almost certainly employed by the Westminster Fire Office, founded 1717, which favoured blue livery plus a badge and firemark consisting of three feathers and a portcullis. The Westminster Fire Office, in fact, insured the Pantheon. Following the fire, it had to pay out £10,000, which represented a quarter of its total assets. Among the directors and members of the Office were some associated with art and design; namely, architects, furniture makers, master builders and sculptors. The Pantheon was

6 J.M.W. Turner, *The Pantheon, Oxford Street, The Morning After the Fire*, (1792). Pencil and watercolour on paper support, 51.6 x 64 cms. © copyright Tate, London 2003, All Rights Reserved

7 Unknown artist, *Fireman of the Westminster Fire Office in Uniform*, (c. 1800). Print after an original picture in the Board Room of the Westminster Fire Office. Longest dimension 27 cms. Print collection of the Guildhall Library, London. © Corporation of London

rebuilt but finally demolished in 1937 to make way for a Marks & Spencer store.

Turner himself had a personal connection to the Pantheon because in 1791 he had painted scenery for its theatre. On the morning after the fire, he spent several hours on site making sketches and then returned to his studio to square them up for a larger watercolour, which was later exhibited at the Royal Academy. He also painted a view of the ruined interior. Hence, his watercolours were based on first-hand observations and can be regarded as examples of pictorial reportage.

Turner's interest in fires continued throughout his life. For instance, in 1834 he depicted a fire that destroyed the Houses of Parliament. In fact, he produced several paintings of this subject and some were distant views that showed no firemen. One gouache and watercolour study did however depict crowds gathered near the blaze and jets of water directed at it indicate the presence of firemen, even though they are hard to discern among all the spectators. Two years later Turner was attracted to another London blaze: *Fire at Fennings Wharf, on the Thames at Bermondsey* (Whitworth Art Gallery, Manchester), which he painted from a viewpoint on the River with boats in the foreground.

## 8 — 1805

## William H. Pyne, Sun Insurance Fireman

This attractive and highly romantic image of a fireman and crew is a coloured aquatint (a method of etching in tone on resin-coated copper plates that results in an image resembling a wash drawing). It appeared as a plate in the book *The Costume of Great Britain* (1805) written and engraved by William H. Pyne (1769–1843). The son of a weaver, he became a writer, engraver, painter and book illustrator. Pyne's volume contained 60 plates (plus commentaries) depicting a range of men and women engaged in different professions and trades. The first edition is now an expensive collector's item but a much cheaper facsimile reprint was issued in 1989.

Despite the book's date, Pyne was depicting the costumes of the previous century. The main figure, a young man, carries a poleaxe in his left hand and holds a lighted link or torch aloft in his right. The torch and the man's wild hair and dynamic pose imply a revolutionary, an incendiary rather than someone who extinguishes fires. He is wearing leather boots and the blue-coloured uniform of the Sun Fire Office with the company's badge on his left sleeve. A badge also appears on the front of his helmet, which with its raised ribs, pointed peak and broad neckpiece resembles those American firemen were to adopt. According to Pyne, the

8 William H. Pyne, *Sun Insurance Fireman*, (1805). Aquatint illustration from *The Costume of Great Britain* (London: William Miller, 1805)

9 Thomas Rowlandson & Augustus Pugin (draughtsmen) & J. Bluck (engraver), *Fire in London* [Albion Mill fire of 1791], (1808). Aquatint/book plate, 26.5 x 22 cms. By permission of the British Library [190.e.1p.35]

'fire-cap' was made from horsehide and strengthened with metal crossbars with an inside made from leather, quilted and stuffed with wool.

Another fireman on the right is directing a branch hosepipe towards a fire that is beyond the right-hand edge of the picture. On the left is a wooden manual pump, probably one designed by Richard Newsham in 1721, being operated by three teams: two teams press levers on either side of the pump while another team on top use their feet to press down on treadles. For this reason, the treadle men were nicknamed 'dancers'. Two decorated leather buckets lie discarded in the left-hand corner and, in the centre, household goods are being salvaged. Since the other firemen have differently coloured attire – green, pink and beige – Pyne was probably indicating the presence of several different fire brigades. In the background, a crowd watches the action and urges the men on with shouts of encouragement. Their presence reminds us that firefighting is a highly visible occupation. Pyne's adjoining commentary is mainly a potted history of fire insurance offices.

## 9 — 1808

### Thomas Rowlandson & Augustus Pugin & J. Bluck, Fire in London

The artist Thomas Rowlandson (1756–1827) is best known for his humorous caricatures, watercolours and prints, executed with a rococo-type line, satirising British society. However, he was also capable of serious works such as *Fire in London* reproduced here. The original was an aquatint published as a plate in *Repository of Arts* (1 September 1808) and then in *Microcosm of London* (London: Rudolph Ackermann, 3 Vols, 1808–11). In fact, Rowlandson drew the figures; Augustus Pugin depicted the architecture; while J. Bluck did the engraving. The event it depicts, however, took place 17 years earlier on 2 March 1791, that is, a major fire at the Albion Mill (or Mills) situated at the south-east end of Blackfriars Bridge, in Southwark. The Bridge, designed by Robert Mylne and built between 1760–68, can be seen to the left while in the centre is the River Thames and on the north bank, set against a night sky, is Wren's St. Paul's Cathedral. One side of the Mill faced the Thames and had a tunnel-like loading bay at water level. On the river is a small craft, which may have been an early fireboat. Another aquatint, by artist John William Edy, shows the Mill on fire from a river viewpoint and depicts such a boat moored close to the bank with hosemen directing a jet of water towards the flames.

The Mill, an elegant four-storey structure completed in 1789, was designed by Samuel Wyatt in a Palladian style; consequently, it did not look particularly industrial even though historians believe it was the first in London to be powered by rotary steam engines, two of which were installed by the Scottish Engineer John Rennie (1761–1821) working for Boulton & Watt. The engines were used to power mills to grind corn to make flour. However, it seems there was a negative public reaction to such technological innovation because the price of bread rose and country millers lost business. At the time, therefore, some welcomed the Mill's destruction and it was rumoured that the fire was due to arson. For instance, a biographer of Rennie has remarked that the Mill was destroyed 'by wilful fire, being obnoxious to popular prejudice, under the mistaken supposition of its being a monopoly'. A contemporary print by Samuel Collings – *Conflagration! Or the Merry Meal Mongers*, published in the *Attic Miscellany* (1 April 1791) – provides evidence of the public mood. It shows a trio of rustic millers celebrating by dancing a jig on Blackfriars Bridge while in the background crowds cheer as the Mill burns.

On the right-hand side of *Fire in London*, flames are leaping from virtually all the Mill's windows and a huge flame and column of smoke emerges from the roof. The fire's light illuminates a wide area. In the foreground, along the bottom edge of the picture, several large fire engines manned by teams of red- and dark blue-jacketed firemen including 'dancers' supply water to hoses whose jets seem too far from the blaze to make much of an impact. (The colours of the uniforms suggests the presence of men from the Phoenix and Sun Fire Offices.) However, two intrepid hosemen have reached a doorway on the extreme right. Also present are salvage men who are busy carrying goods to safety. More horse-drawn engines and wagons are arriving at speed across the bridge plus a stream of curious onlookers; a few soldiers are attempting to hold the latter back. In short, Rowlandson and Pugin generated an animated if depoliticised composition, which even today communicates a strong sense of the power of fire and the agitation and excitement involved in combating and observing it.

Despite the strenuous efforts of the fire companies, they were unable to prevent the Mill's total destruction. This is borne out by a pen-and-wash drawing by the artist James Moore and a watercolour by the artist Thomas Girtin, produced soon after the fire, which depict smouldering ruins. The Albion Mill fire proved to be a popular subject with artists and print makers between 1791 and 1808. Brian Maidment, a British print historian, discusses no less than 15 examples in his book *Reading Popular Prints, 1790–1870*, (2001).

10 Robert Cruikshank, The *'House of Accommodation' in Flames* ..., (1821). Aquatint/bookplate published in *Life in London* (1821)

## 10 1821

## Robert Cruikshank, The 'House of Accommodation' in Flames

In 1821, Sherwood, Neely & Jones published *Life in London; or, the Day and Night Scenes of Jerry Hawthorn Esq., and his Elegant Friend Corinthian Tom, Accompanied by Bob Logic, the Oxonian, in their Rambles and Sprees through the Metropolis.* Written by the journalist and sportswriter Pierce Egan, the elder (1772–1849), it recounts in 36 scenes the adventures of fictional young 'men about town' during the Regency period (1810–20).

Hand-coloured engravings illustrated the book. One represents a hectic street scene in which firemen, dressed in red jackets and black stove-pipe hats, operate a hand pump, direct hoses, wield axes and climb ladders to fight a fire and rescue young women from a 'House of Accommodation'. Paupers lived in such houses but in this instance, the description seems to have been a euphemism for a brothel. The engraving's caption reads: 'The "House of Accommodation" in flames! – The inmates put to the flight. JERRY narrowly escapes with his life but preserves ELLEN PRETTYFLOWER his paramour from an untimely death.' There is danger and drama in this print but also humour.

Smoke from the fire pours into the street, which is crowded with figures moving in all directions and so a sense of alarm and confusion is communicated. One sympathises with the firemen trying to perform their duties in the midst of such chaos. On the left, a fashionable middle-class couple arrive to observe events. In the centre, a strong fellow (Jerry?) carries a swooning damsel (Ellen?) in his arms. Beside him, a salvage man holds a pair of boots and balances a large sack of goods on his head. Nearby, a hoseman directs a jet of water close to a woman in a green dress high up a ladder. In the right-hand corner, a nightwatchman holds a lantern in his left hand and spins a rattle with his right. Immediately in front of him, several men hold a sheet to catch a victim falling from the burning building. A repertoire of firefighting techniques is thus depicted.

Isaac Robert Cruikshank (1789–1856, known as 'Robert Cruikshank'), the English artist responsible, came from a family of artists. His grandfather had been a professional artist and his father Isaac (1756–1811) was a noted caricaturist and watercolourist, and his younger brother George (1792–1878) was a prolific illustrator. (His son Percy also became an artist.) George was to become the most famous and highly regarded of the Cruikshanks. He too contributed illustrations to *Life in London* but then withdrew from the project on moral grounds: it seems he objected to the lack of condemnation of vice in Egan's tale. According to Egan, the character of Bob Logic was based on himself, while Tom was based on Robert and Jerry on George. The book proved to be highly popular and was adapted for the theatre and reprinted many times. In the twentieth century, Hollywood animators employed the names 'Tom and Jerry' for the cartoon characters Tom (a cat) and Jerry (a mouse).

## 11 1825

## Unknown artisan, Protector Fire Insurance Company [firemark]

## 12 1827

## R. Smirke & J. Thompson, Protector Fire Insurance Company

## 13 1805–67

## Unknown artisan, Firemark of the Birmingham Fire Office

## 14 1848

## Unknown artisan, Firemark of the Associated Firemen's Insurance Company, Baltimore, Maryland

Firemarks and plates were small signs devised by British and American fire insurance companies for nailing to the external walls of buildings that they insured. The signs were variously shaped and made from carved wood or metals such as copper, brass, cast iron and lead, and had

11 Unknown artisan, *Protector Fire Insurance Company* [firemark], (1825). Relief sculpture, copper 22 x 24 cms. York Castle Museum collection. Photo by Richard Stansfield © York Castle Museum 2003

12 R. Smirke (painter) & J. Thompson (engraver), *Protector Fire Insurance Company* [image appearing on a certificate dated 1827]. Photo courtesy of Jarndyce Antiquarian Booksellers

crude images, lettering and numbers on them in shallow relief. Some figures or animals were designed as cut-outs but others were set against backgrounds. The marks, which were often brightly gilded or coloured, were a simple visual method of informing British insurance fire brigades which properties they were to save. (In the United States, volunteer fire companies predated insurance companies and American firemen were willing to fight fires even if the buildings were uninsured.) Of course, as street signs, the marks also functioned as advertisements for the insurance companies; in addition, the designs appeared on arm badges or brassards – some made from silver – worn by their firemen.

One might have expected firemarks to employ images of firemen but most British ones do not. Two exceptions can be cited: first, that of the Protector Fire Insurance Company of London (1825–35), which has an image of a spindly fireman using a branch hosepipe to douse a blaze. In this instance, he wears a low top hat, probably made from beaver, rather than a leather helmet. Red-coloured flames and black smoke are rendered in a stylised fashion. In the background is old Westminster Bridge and the design derives from a painting by Robert Smirke (1752–1845), a painter, book illustrator and member of the Royal Academy, who was also one of the insurance company's auditors. Second, the metal plate of the Birmingham Fire Office founded in 1805, which features a fireman holding an axe with his right hand standing in front of a manual engine. Six variations of this plate were produced and made from copper, brass and iron. A branch office in Leamington Spa also had a painted metal sculpture of a pump flanked by two firemen wearing green coats, blue trousers and black boots above its porch.

In the United States, more firemarks depict firemen. For example, in 1848 the Associated Firemen's Insurance Company of Baltimore (1847–99) issued an oval-shaped sign made from cast iron that depicted a fireman running from right to left. He wears a cape, holds a burning brand in one hand and is using a speaking trumpet to shout the word 'Fire'. (Such trumpets were used by foremen or assistant foremen, who were elected annually by their fellows, to sound alarms and shout orders. Highly decorated ones were also produced as presentation items.) Clearly, he is sounding the alarm and leading the way to the fire.

A similar mark, also made from cast iron, was issued in 1851 by the Associated Firemen's Assurance Company of Pittsburgh (1850–1930). The fireman is again wearing a cape and shouting into a trumpet but

13 (left) Unknown artisan, *Firemark of the Birmingham Fire Office*, (1805–67). Relief sculpture, copper 36 x 30.5 cms. Private collection

14 (right) Unknown artisan, *Firemark of the Associated Firemen's Insurance Company*, Baltimore, Maryland, (1848). Relief sculpture, cast iron, 30 x 25 cms Private collection

this time carries a wrench for tapping a nearby fire hydrant. The Home Insurance Company, New Haven, Connecticut and the People's Insurance Company, New Orleans, Louisiana issued additional examples. A number of firemarks also depict manual fire engines.

Firemarks possess historical and curiosity values, and the aesthetic appeal of folk art. Reference books and websites catalogue them and they have been keenly collected since the nineteenth century, so that now hardly any of the originals remain on the façades of buildings. However, such companies as Wakerley Perkins of England and Heritage Metalcraft of Maine manufacture sand-moulded, hand-painted, aluminium replicas, which are bought and displayed as decoration. Heritage Metalcraft has also issued a new plaque commemorating 9/11: it features a Maltese Cross enclosing the initials of the New York Fire Department plus the phrase 'Lest We Forget'.

## 15 — 1830

## Unknown artist, Portrait of George Mead

## 16 — 1832

## Unknown artist, Royal Exchange Fire Office liveried Fireman

The bust-length portrait of a fireman by an unknown British artist, which dates from the early nineteenth century, is thought to be of George Mead, the foreman or captain of the fire brigade of the Guardian Fire & Life Assurance Company from 1822 to 1833. (Foremen and deputy foremen were full-time employees while the men under them tended to be part-timers who were paid fees for being on call, attending drills and fires.) Red and gold trimmings, medallion-like buttons and an ornate brassard enliven his dark brown uniform. It is so expensive and elaborate as to be impractical for fire-fighting and so it was probably a dress or parade uniform. Although the pose is rather formal, the portrait is competently executed.

Mead was clearly a handsome but heavy-set individual with no neck (his head merges seamlessly with his body); consequently, it is hard to imagine him taking an active part in fighting fires alongside his men. However, wearing this colourful costume, Mead would have been a walking advertisement for his company's fire insurance policies and its brigade. It seems such fine clothes ensured there was no shortage of applicants for the job.

The Guardian Fire & Life Assurance Company was founded in London in 1821 and its brassard featured a silver cast and chased figure of Athene, the Greek goddess of arts, industries, prosperity and prudent warfare. She stands among clouds and holds a spear in her right hand and an olive branch in her left. Above her, there is a ribbon bearing the name of the company and all are contained within a leaf and acorn motif border. Mead's badge in the portrait is numbered 1. A surviving badge, numbered 2, dated 1821, bears the mark of Philip Venner Firmin (c. 1822–52) of the Strand, London who made sets of badges for various insurance company fire brigades.

Christie's, the auction house, held a sale of items from the Royal Exchange Collection on 26 June 2001 at their South Kensington saleroom. (The Guardian and Royal Exchange Companies merged in 1968.) Mead's

15 Unknown artist, *Portrait of George Mead*, (c. 1830). Painting, oil on canvas, 76.2 x 62.9 cms. Photo © Christie's Images Ltd, London

16 Unknown artist, *Royal Exchange Fire Office liveried Fireman*, (1832). Print 14 x 18 cms. (after a painting? Size and location unknown).

portrait was described as 'English Provincial School circa 1830' and valued at between £2000 and £3000.

Another early, nineteenth-century painting depicts a fireman in a proud but stiff pose. He is wearing the pea-green plush suit with tunic buttons, a beaver hat with gold-lace band and white stockings uniform of the Royal Exchange Fire Office of Cornhill and Pall Mall. An earlier version of the uniform had been coloured yellow and lined with pink, but this had proved impractical. In London, rival fire brigades had differently coloured uniforms in order to distinguish them. One would not perhaps realise that this well-dressed individual is a fireman except for fact that he is holding a felling axe and has a large, oval-shaped company badge on his left sleeve. The firemark of the Royal Exchange Assurance, established in 1720, featured an image of the Royal Exchange building surmounted by a crown. The first building occupied by the Royal Exchange dated from 1570 but was destroyed by the Great Fire of 1666. The second building – which appeared on the firemark – was designed by Edward Jarman and dated from 1669. Located in Cornhill, this imposing neo-classical structure with arcades and a distinctive clock tower was also destroyed by a fire in 1838. One hopes the Royal Exchange company insured their own buildings!

The man's uniform is not functional in the modern sense and appears to be that of a foreman or one used for ceremonial occasions such as annual parades rather than for firefighting itself. Like Mead's, the man's figure is portly rather than athletic. Behind him to the right is wooden manual fire pump, probably a Newsham-type made by Bristow's, which has also been coloured green and decorated with the company's firemark. To the left is a classical-style balustrade with steps leading down to the River Thames, which of course was a major source of water to extinguish fires occurring along its banks. It may also be allusion to the origin of many of London's firemen who were recruited from the ranks of the river's watermen or ferrymen.

While this painting is no masterpiece, it is a professionally executed and dignified portrait. It was probably commissioned to adorn the Royal Exchange headquarters and has been reproduced in most histories of firefighting in Britain and on a cigarette card issued by John Player & Sons in 1930.

A third portrait by an unknown artist of yet another portly fireman named Robbins exists. The Norwich Union Fire Insurance Society, founded in 1797, employed him and supplied his uniform of dark-green frock coat plus red breeches and waistcoat. The oval-shaped badge on his arm depicts a seated allegorical female figure of Justice blindfold. In composition and style, this oil painting is very similar to the Mead portrait.

Portraits such as these established a tradition that was to be continued in the portraits, both painted and photographic, of fire chiefs in Australia, Britain, Europe and the United States.

## 17 — 1830

## James Pollard & R.G. Reeve, London Fire Engines: The Noble Protectors of Lives and Property

## 18 — 1866

## John Cameron, The Life of a Fireman. The Metropolitan System

At first, pumps on wheels were pulled by men, then by horses, and then propelled by steam power and, finally, in the early twentieth century, by petrol motors. As Pollard's design for an aquatint indicates, galloping horses pulling wagons with fire engines on board became a popular subject for artists from the 1830s onwards. Initially, artists favoured the side view with vehicles moving across the picture plane from right-to-left or left-to-right, but later they used three-quarter and head-on viewpoints.

Pollard (1792–1867) was a prolific British artist who created scenes of mail coaches, fishing, fox hunting and urban thoroughfares. In his aquatint, no less than three

17 James Pollard (artist) & R.G. Reeve (engraver), *London Fire Engines: The Noble Protectors of Lives and Property*, (c. 1830). Aquatint on paper, 78 cms across. Published by Thomas McLean of London. Guildhall Library Print Room, Wakefield Collection © Corporation of London

18 John Cameron, *The Life of a Fireman. The Metropolitan System*, (1866). Hand-coloured lithograph published by Currier & Ives, New York, 42.7 x 66.4 cms. Library of Congress, Prints and Photographs division

horse-drawn engines belonging to different fire offices – the County, Phoenix and Westminster – are shown racing one another – like the chariots of ancient Rome – along a city street to reach a night-time fire. (Of course, once an engine reached a fire, the horses had to be moved to a place of safety.) Firemen on board the engines, it seems, used to shout 'Hi, Hi, Hi!' to clear the way ahead. At the fire itself, a manual-pump is already in operation and, on the far pavement, a team of men urged on by a beadle are dragging the parish-pump. Behind them goods are being salvaged and nearby, two men – one a fireman – are using a tool to turn a stopcock to access a water supply beneath the street. In the bottom left-hand corner, there is the familiar figure of a night-watchman carrying a lamp and a rattle.

Aroused by all the commotion, local residents have thrown open their windows and in the street below there are excited adults and children hurrying towards the light of the flames. Paradoxically, the scene has almost a festive air. The printmaker's use of the description 'noble protectors of lives and property' in the title may be taken as referring to London's firemen as much as to fire engines and therefore is an early indication of the popular conception of firemen as noble heroes.

Images of similar scenes with titles such as 'going to the fire' or 'the fire run' also became popular in the United States. For example, in 1866, Currier & Ives published a print entitled *The Life of a Fireman. The Metropolitan System*, showing side views of a steamer, a carriage hose reel and a ladder carriage each pulled by pairs of horses. 'Metropolitan System' was a reference to the new professional fire department that had replaced New York's volunteer department in 1865. John Cameron, the artist, was active in New York from 1848 to 1872 and received several commissions from Currier & Ives. As the image reveals, fire vehicles at that time were not designed to transport all the members of a company and so some had to run alongside. One presumes that many must have arrived at fires exhausted.

## 19 — 1836

## George Cruikshank, The Parish Engine

George Cruikshank (1792–1878) came from a family of artists and was one of the most talented and prolific British illustrators and caricaturists of the nineteenth century. During 1835 and 1836, he collaborated with the soon-to-be-famous writer Charles Dickens (1812–70) who employed the pen name 'Boz' when submitting 'sketches' of London life to several magazines from 1832 onwards. Cruikshank provided illustrations for the latter's first book, a collection entitled: *Sketches by 'Boz', Illustrative of Every-day Life and Every-day People*, (London: John Macrone, 2 Vols 1836).

19 George Cruikshank, *The Parish Engine*, (1836). Etching on steel plate/book illustration

In 1839, a second expanded edition was issued and Cruikshank reworked and enlarged most of his original subjects and added 13 new ones.

Dickens' description and Cruikshank's depiction of the beadle, a character of the period responsible for local property who wore a cocked hat, a 'state' coat and carried a staff of authority, and the parish's manual fire engine pushed and dragged by a mob of unruly urchins are both comical. (According to William H. Pyne, 'marshalling charity children' was one of the beadle's duties.) Following an Act passed in 1772 during the reign of George III, London parishes and their churchwardens were supposed to acquire, maintain and store fire engines (mostly in churchyards or workhouses). 106 parishes had one but 104 did not. At large fires, parish-pumps were virtually useless. Dickens must have been aware of the patchy and poor protection against fire provided by the parish-pump system and this is why he satirised it.

While curious neighbours gaze skywards, the portly and pompous local official knocks on the door of a house smelling strongly of soot, which displays an optician's sign and a firemark of the Sun Fire Office. It seems the parish-pump was called out after two boys reported a chimney fire but, in Dickens' account, no one answers the door. Next day the beadle will visit again to demand a call-out charge.

Dickens also describes how he once witnessed a fire at which a beadle and 18 boys operated a pump for 20 minutes without effect because they had not realised its cistern had first to be filled with water via buckets. Evidently, neither Dickens nor Cruikshank had much respect for local firefighting arrangements, which were clearly amateurish compared to the brigades employed by the insurance companies.

## 20 1836

### Unknown artist, Chief Director of Mechanic Fire Company No. 1, Louisville

This larger-than-life wooden figure of a Chief Director (aka 'General Shaw') holding a trumpet and pointing towards a fire by an unknown artist or artisan is thought to date from the 1830s. It seems the volunteer fire companies of Louisville, Kentucky had such ranks as Chief Director and Directors of engine, pipe and line. The Mechanic Fire Company No. 1 was incorporated in 1834 and red helmets and vermilion coats identified its members.

The figure has the crude characteristics and charm of folk art. During the eighteenth and nineteenth centuries, ships' carpenters often carved figureheads to ornament the prows of sailing ships and local artisans carved free-standing trade figures for promotional purposes outside shops. 'Cigar Store Indians' were common examples of the latter. These effigies, distant relatives of the cult statues of antiquity, were 'place- and product-identified'. Many examples have been lost because display out of doors resulted in decay but those that have survived are generally preserved in collections of American folk art.

The figure's surface now possesses a rich patina. Originally, it must have looked hyperrealistic because paint samples taken during conservation suggest that the face and hands were flesh-coloured and that the flaring, double-breasted surcoat had gold buttons. The red colouration of the helmet's brim can still be seen. The Director has rather stern features and long sideburns. The absence of idealisation implies this was a portrait of a real individual.

Such figures were often displayed out of doors on the façades or roofs of firehouses or buildings occupied by firemen's associations. Where the Louisville figure was originally situated is not known but it was probably at the top of a firehouse tower because the figure had a pivot so that it could be rotated from below to point in the direction of any blaze. What is known is that from 1902 to 1922, the figure stood beneath a cupola on the roof of the Volunteer Fireman's Historical Association building (an ex-fire station) at 313 South First Street, Louisville. Two similar wooden figures will be discussed later.

20 Unknown artist, *Chief Director of Mechanic Fire Company No. 1, Louisville*, (c. 1830s). Sculpture, carved and painted wood, 198 cms high. Louisville, Kentucky: J.B. Speed Art Museum collection. Gift of William C. Bittner, 1949. Photo: Moseley Photography, courtesy of the Museum

## 21 1841

### Christian Mayr, Officers of the Volunteer Fire Department

In the history of art, the group portrait reached a peak of achievement in seventeenth-century Holland, then a successful, commercial, bourgeois society, when artists such as Frans Hals and Rembrandt were commissioned to record the officers of civic militia companies and the regents of hospitals. Christian Friedrich Mayr's attractive frieze-like painting of 33 volunteer fire officers from Charleston, a South Carolina seaport founded in 1670, continues the Dutch group portrait tradition. All the officers, of course, were white because South Carolina was then a slave state.

Mayr was born in Nuremberg, Germany in 1803. His father Johann von Mayer was an artist and his son studied fine art at academies in Nuremberg and Munich before departing for the United States in about 1834. Mayr became a successful genre, portrait and history painter catering for upper middle class Americans. In 1838, he became an American citizen in Charleston where he worked until 1843. He moved to New York and became a member of the National Academy of Design. He died there ten years later.

21 Christian Mayr, *Officers of the Volunteer Fire Department*, (1841). Painting, oil on canvas, 114.3 x 166.4 cms. Reproduced courtesy of Collection of City Hall, Charleston, South Carolina

Mayr's group portrait depicts the male social elite of the city in their roles as fire service engineers, presidents, foremen, superintendents and fire masters, plus local dignitaries such as the German-born Mayor Jacob F. Mintzing. The officers' diverse uniforms and dress coats, some military in style, some white, some black and some brown, indicate that they belonged to different fire companies; namely, the Aetna, Axemen, Charleston Neck, Marion, Palmetto, Phoenix, Vigilant, and the German. The name of the latter reveals that Charleston had a significant number of immigrants from Germany and this was probably one reason why Mayr was attracted to the city. (A German Friendly Society to assist new German immigrants had been established in 1766.) Mayr, it seems, was a member of the Deutsche Feuer Compagnie, which was established in 1838 following a major blaze in Charleston, and he included his own portrait in the painting.

The figures stand informally in the middle of Meeting Street in the centre of Charleston in readiness to form a procession. All rivalry between the fire companies has been suspended as they assemble for a joint celebration. Most figures gaze directly at the viewer. While some wear fire hats, others wear top hats; most hold speaking trumpets but a couple have bugles and several grip wooden staffs capped with flame-like metal ornaments. (Since colonial times, fire wardens had carried such staffs as symbols of their authority.) Two dogs relax on the ground in the centre of the composition. In the background can be seen the City Hall, the Fireproof Building and the imposing spire of the Circular Church.

Mayr's composition conveys a harmonious impression because the rectangular shape of his canvas accords with the 'divine proportion' of the Golden Section, that is, the ratio 1 (vertical) to 1.618... (horizontal). One of the vertical bisecting lines in a geometric rendition of the Golden Section coincides with the staff held by the tallest figure off centre to the right.

In August 1841, Mayr displayed his completed canvas in his rooms and charged visitors an entrance fee. He also sold raffle tickets for a dollar each with the painting as the prize. The winner was a son of Richard Wainwright Bacot, President of the Phoenix Fire Company, whose widow gifted the work to the City of Charleston in 1881. A diagram identifying each individual in the painting has been displayed alongside the painting in City Hall.

## 22 1848

### James Brown, F.S. Chanfrau in the character of 'Mose'

## 23 2002

### Mary Pope Osborne & Steve Johnson/ Lou Fancher, Mose in action

Stage plays were another means by which the popular cult of the fireman was fostered during the nineteenth century. On 15 February 1848, at the Olympic Theatre, New York, a new play entitled *A Glance at New York, A Local Drama in Two Acts* by Benjamin A. Baker proved to be a hit. It starred the comic actor and volunteer fireman Frank S. Chanfrau (1824–84) in the role of a New York fire laddie called 'Mose'. The play toured the United States and sequels were produced during the 1850s (known as 'Mosaics') and so Mose rapidly became an American folk hero.

Several lithographs were published at the time depicting Chanfrau as Mose wearing a 'plug' or stovepipe hat, boots and a red fireman's shirt. One showed him with 'soap-locks' (hair plastered with soap) standing in a defiant mood (probably he had just been reprimanded for insubordination) with his back to a fire engine and crew hard at work, while a second showed him holding a basket with a baby inside. E. & J. Brown published the former, by James Brown (an artist active in New York 1844–58), while J. Baillie published the latter and John L. Magee, an illustrator and political cartoonist, was the creator. Another lithograph by Brown depicted Mose fighting with other firemen. Other prints show Mose comforting a crying street girl, courting a young woman, sitting on the curb, and treading on a hosepipe to cut off a rival fire company's water supply.

Mose appealed because he was a butcher of humble origin, a strong, irreverent, fun-loving, street-smart Bowery B'hoy or dandy who loved to boast and fight (against landlords, politicians and thieves), consume beer and smoke cigars, and embrace women. However, he was also a keen and kindly firefighter who rescued victims of fire and helped country bumpkins. On stage, he was over six-feet tall and had flaming red hair. His aggressive dialect – 'Get off dem hose or I'll hit yer with my spanner' – was much imitated by his juvenile fans. However, some real firemen disliked Mose because they considered the rowdy violent character inaccurate and degrading, even though the truth was that rivalries between volunteer companies often resulted in practical jokes and street brawls.

Although Baker's play was a melodrama, it was rooted in the reality of New York life and indeed Mose was based on a real Irish-American fire laddie called Mose Humphreys (aka 'Old' or 'Big Mose'), who was foreman of the Lady Washington Engine Company No. 40, based in Mulberry Street. He was a printer or typesetter rather than a butcher but is reported to have had long arms and huge hands, and to have risked his life to save women and children from burning buildings. Chanfrau had once witnessed a fight between his elder brother Henry (a rival fireman) and Mose, which Henry won. Mose soon left New York for the gold fields of California and ended up in Honolulu.

The appeal of the fictional Mose declined in the 1860s and 1870s as volunteer firemen were increasingly replaced by better behaved, paid professionals. However, his influence as an American hero is still felt – witness the range of products in which 'Big Mose' has featured since the 1950s: illustrated books by Katherine Binney Shippen (1953), Harold W. Felton (1955, 1969), and Mary Pope Osborne (2002); an animated video with commentary by Michael Keaton marketed in 1993 by Macmillan/McGraw Hill in the Story Lane Theater series; a camp play starring Zorikh Lequidre mounted in Houston, Texas, in 2000; a musical composition by John Deak performed in New York in 2002;

22 James Brown, *F.S. Chanfrau in the character of 'Mose'*, (1848). Lithograph, 32 x 24 cms. Published by E & J. Brown, New York. Washington DC: Smithsonian Institution, National Museum of American History, Peters Collection. Photo 600446–B. Picture caption reads: 'I'm bound not to run wid der machine any more.'

23 Mary Pope Osborne (writer) & Steve Johnson/Lou Fancher (illustrators), *Mose in action, an illustration from New York's Bravest*, (New York: Alfred A. Knopf, 2002). Reproduced courtesy of Random House

and a revival of *A Glance at New York* by director Randy Sharp and the off-Broadway company Axis (May to July 2003).

In Osborne's book for children illustrated by Steve Johnson and Lou Fancher, Mose has made the transition from folk hero to superhero: Osborne claims he is eight-feet-tall and capable of lifting and carrying streetcars full of people, rescuing babies in his stovepipe hat, and so forth!

## 24 1848

### Robert E. Launitz, Firemen's Monument

Public memorials to fallen firefighters, which have sculpted figures in bronze, zinc, marble and wood, are now commonplace across the United States but they have only been in existence for about 150 years: they date from the funerary art of the second half of the nineteenth century. (While similar memorials are to be found outside the United States, no other country has so many and is adding so many new ones.) Most are sited in peaceful cemeteries where fallen firefighters are buried, while others are located outside firehouses, courthouses and libraries or in streets or public parks. Some are incorporated into drinking fountains, which can probably be explained by the need to water the horses that pulled fire engines up until the 1920s and by the fact that firefighters employ water to extinguish fires.

Typically, in early examples, a single figure perches on top of a plinth or column. While the height of these memorials varies, the firemen are normally raised above the viewer's head as a sign of their importance and nearness to heaven. Height also enables the memorial to be visible from a distance and protects the sculpture from vandalism. The disadvantage, in the case of tall memorials, is that the sculptures on top cannot be seen clearly. Sculptures located out of doors, of course, are subject to weathering that gradually discolours and erodes them and so some were given coats of paint to protect them and to make them appear more life-like.

Figures pose on plinths with solemn dignity: most nineteenth-century sculptors made a virtue of stillness and did not try to suggest action or movement apart perhaps from one leg stepping forward. (This pose dates back to the Kouros – 'young man' – statues of ancient Greece.) Some figures wear uniforms, helmets and hold speaking trumpets or hosepipes, while others carry a coat or a small child in one arm and a lantern in the other. Many figures were carved from white marble. As Thomas B. Brumbaugh has remarked, 'There was a strongly held notion in the aesthetics of the time, elaborated by such writers as Emerson and Hawthorne, that much of the quality of a work of art lay in an intrinsic morality of the substance from which it was made. White marble represented the perfect transcendental material for an age of monumental rhetoric.' Some monuments also have stone or bronze bas-reliefs on their pedestals depicting firefighting scenes such as horse-drawn engines answering alarms.

Most of the sculptors responsible were professional fine artists trained in the studios or workshops of other American or European artists. Many were immigrants from Europe. Besides carving or modelling firemen, they undertook commissions for Civil War monuments and statues of American politicians. Other artists are little known or unknown because they were artisans employed by local marble companies or bronze foundries or iron works. (When the Civil War ended, some foundries that had made swords and cannons switched to casting metal sculptures.) Normally, the funds needed to pay for the sculptures were raised by public subscription, by fire departments and firemen's benevolent associations or were donated by wealthy individuals.

24 Robert E. Launitz, *Firemen's Monument*, (1848). Memorial, marble and granite. Brooklyn: Green-Wood Cemetery. Photo: Jeffrey L. Richman

One of the earliest American memorials of this kind is the *Firemen's Monument* (1848) erected on Tulip Hill in Green-Wood Cemetery, Brooklyn, which was founded in 1838 to serve the city of New York. On a cube-shaped pedestal of granite, a pyramidal marble column rises high into the air. On top stands a marble fireman who holds a trumpet in his right hand and cradles a small child in his left. A nineteenth-century author commented: 'The attitude is spirited, and the general effect very good.' (Nearby are similar monuments commemorating firemen Henry Fargis and George Kerr.) Panels on the base list the names of the fallen and it is decorated with carvings of firefighting accoutrements and tools. The monument was commissioned by the Fire Department of New York 'in memory of their companions who perished in discharge of their duty'.

The sculptor responsible was Robert E. Launitz (aka Robert Eberhard Schmidt von der Launitz, 1806–70) who was born in Riga, Latvia into an aristocratic family. He studied sculpture in Rome and then spent four years as an apprentice to the Danish neo-classicist Bertel Thorwaldsen. In 1828, he came to New York and honed his skills in the marble yard of John Frazee. Often he provided designs or plaster models of sculptures, which were then carved by others, sometimes by artisans in Italy. After completing many memorials during a long career, Launitz was dubbed 'Father of monumental art in America'.

Many firemen's monuments became better known by being depicted on picture postcards, which are still being sold via auction sites on the World Wide Web, and by small-scale models issued in editions by contemporary giftware and collectibles manufacturers. Dozens of designs exist of firefighter sculptures in the form of small figurines serving ornamental purposes, but in the main their aesthetic quality is poor.

## 25 — 1850

### Charles Washington Wright, Portrait of Franklin as a Fireman

Benjamin Franklin (1706–90), a leading eighteenth-century American politician, was also famous for his scientific achievements and inventions, printings and writings. In addition, he is remembered as a volunteer fireman, as indeed the founder of the first American fire company, the Union (1736–1820), in Philadelphia. It was a mutual assistance association with 30 members, each of whom had to pay for six leather buckets and two linen salvage bags. (When most companies consisted of volunteers, they often included prominent citizens.) Franklin also helped to found the first American fire insurance company named the 'Contributionship' (1752– ). Its firemark consisted of four hands clasped together at the wrists and so it was also known as the 'Hand-in-Hand' company.

25 Charles Washington Wright, *Portrait of Franklin as a Fireman*, (c. 1850). Painting, oil on canvas, 75 x 62 cms. Photo: Philadelphia, PA: CIGNA Museum & Art Collection

The oil portrait showing Franklin as a fire chief aged about 40, has been ascribed to Charles Washington Wright (1824–69) who was active in New York from 1844. Wright was a medallist and wood engraver who undertook illustration commissions for such magazines as *Harper's Weekly* and *Frank Leslie's Illustrated Weekly*. Since the portrait dates from around 1850, it was produced long after Franklin's death. Presumably, it was based on an earlier likeness. The white hat with a rendering of a wooden pump and the words 'Union Fire Company' was of a type only introduced in the 1830s and the Union Fire House building (which never actually existed) in the background were later additions made around 1900 at the behest of one of Franklin's descendants; consequently, the portrait is a curious temporal confection. Most viewers sense something is wrong with the portrait because the helmet rests so awkwardly on Franklin's head.

In 1948, a reverse image of the painting was reproduced on an art cover envelope to honour the volunteer firemen of America. In Philadelphia there are several public sculptures of Franklin and one in Fourth Street, created by Dexter Jones (1926–86) in 1968, is a bas-relief set against a brick wall of his head seen in profile wearing a firefighter's helmet. Some small children in the United States are introduced to the famous man via a soft toy entitled the Franklin Fireman [Teddy] Bear, sold by the Franklin Institute of Philadelphia.

## 26 — 1850

### W. Turner, William M. Tweed

## 27 — 1897

### Thomas Nast, Hurrah Boys, We are 100 years old

Turner's engraved portrait of William M. Tweed as Foreman of the Americus Engine Company No. 6 of New York may have been based on a painting by fireman-artist Joseph Hoffman Johnson because the latter was an Assistant Foreman under Tweed in the same company. Tweed has a fixed gaze and stands as rigidly as a wooden statue. He stretches one of his braces with his right thumb and his belly already shows signs of the corpulence that would mark his later years. Tweed was born in 1823 in New York into a family of Scottish origin. Initially, he ran a chair-making business.

As a teenager, he was attracted by the excitement and sociability of firefighting and joined several volunteer companies. Later, in 1849, he helped to organise a new one called Americus (aka 'Big Six' and 'Tiger'), whose John Agnew, Philadelphia-style pumper had panels painted by Johnson including one with a snarling Bengal tiger's head. (Subsequently, the tiger image became associated with Tweed and was employed by political cartoonists as a symbol of Tammany Hall corruption.) When Tweed was elected its foreman in May 1850, he was presented with a speaking trumpet (the one under his right arm in the illustration) and a watch. However, he was soon temporarily expelled for encouraging his men to attack another fire company.

Tweed was one of many ambitious Americans who used the electoral muscle of fire companies to gain political power in local and state government. He held the posts of alderman, commissioner of public works, fire commissioner and senator, but he also became notorious for bribery, corruption, greed and theft. (One of his biographers defends Tweed against these charges.) When Tweed's Democrat political machine 'Tammany Hall' or 'The Tweed Ring' controlled New York, he acquired the nickname 'Boss'. Ironically, he died in 1878 in a prison he had helped to commission. Tweed remains one of the most famous names of nineteenth-century America and even appears in Martin Scorsese's epic movie *Gangs of New York* (2002). The British character actor Jim Broadbent (b. 1949, Lincolnshire) played the role. In addition, this film has scenes of rival fire companies fighting one another while a building burns and is being ransacked by thieves.

Surprisingly, a graphic artist was instrumental in Tweed's downfall, namely, Thomas Nast (1840–1902), dubbed 'the father of the American political cartoon'. Born in Germany, Nast was taken to the United States as a child in 1846. In 1854, Nast studied art with Theodore Kaufmann and a year later, he worked as a copyist at the Thomas Jefferson Bryant Gallery of Christian Art. Nast later became a watercolour painter, a muralist, a Civil War reporter, an American ambassador, a symbol inventor and a prolific caricaturist and illustrator for books and magazines such as the *New York Illustrated News* and *Harper's Weekly*. A radical Republican, he produced many satirical cartoons during the period 1868–71 attacking Tweed who once remarked that he was not bothered by written attacks because his constituents could not read, but Nast's images were a threat because even the illiterate could understand them.

Although Tweed and Nast were different in many respects, they both wore beards and were volunteer firemen. Nast lived in Morristown, New Jersey for 30 years

26 W. Turner, *William M. Tweed*, (c. 1850). Engraving/illustration in A. Costello's *Our Firemen*, (1887)

27 Thomas Nast, *Hurrah Boys, We are 100 years old*, (1897). Oil painting on canvas, 102.5 x 153 cms. Morristown, New Jersey: Macculloch Hall Historical Museum. Photo courtesy of the Independent Hose Co. No. 1

and was an honorary member of its Independent Hose Company during the 1890s. Firefighting in Morristown dated back to 1797 and so, on its 100th anniversary, Nast painted a humorous, luridly coloured self-portrait in fireman's attire (helmet, red shirt, dark trousers). Nast's depiction of himself as a cheerful rotund figure recalls one of his most famous creations: a cherubic, white-bearded Santa Claus. The painting incorporates a caption proclaiming: 'Hurrah Boys, we are 100 years old.'

The Independent Hose Company, founded in 1834, whose members were mostly Protestants, occupied half of a brick firehouse at 15 Market Street. The Washington Hose Company, whose members were mostly Catholics, occupied the other half but such was the rivalry between the two that there was no connecting door. In 1870, a gang of incendiaries plagued Morristown. When the police caught the gang, they discovered its ringleader was a saloon-keeper and the foreman of the Washington Hose Company! It seems he set fires on Saturday nights in order to earn money from men thirsty from firefighting.

## 28 — 1850

## Unknown artist, The Fireman: The Nobleman of our Republic

## 29 — 1850

## Unknown artist, Up to the Bend

The American print entitled *The Fireman*, one of a set of four executed by an unknown but professional artist, is a particularly vivid depiction a volunteer firefighting company, attired in red shirts and blue trousers, working the levers or brakes of a side-stroke, gooseneck, hand pump. The normal speed of operation was 60 strokes per minute but, if necessary, could be much higher. Standing on the engine is a foreman who uses a speaking trumpet to bark orders and words of encouragement. In the background, there is a house on fire and a fireman with a hosepipe climbing a ladder. The outside of the air chamber of the pump is richly decorated with ornamentation and a panel painting. The print's subtitle proudly asserts that, in the democratic republic of the United States, firemen constituted its nobility.

*Up to the Bend*, in contrast, is a painting that has been executed by an amateur or unsophisticated artist. Jean Dubuffet, a French painter, coined the term '*l'art brut*' to characterise art that was raw and uncultured. He admired its energy, innocence and originality. The animated scene commemorates an actual event that occurred in Mackerelville, a Manhattan shanty town, where many Irish immigrants lived. (Augustine E. Costello gives a full account in his book *Our Firemen* [1887] but fails to give the exact date, which must have been before 1852.) In the crowded foreground, two rival fire companies battle to subdue a night-time fire consuming wooden houses. On the left, Howard Engine

28 Unknown artist, *The Fireman: The Nobleman of our Republic*, (c. 1850). Print published by T.W. Strong of New York. Photo source: The Harold V. Smith Collection, Home Insurance Company, New York.

29 Unknown artist, *Up to the Bend*, (c. 1850). Painting on linoleum. Photo source: The Harold V. Smith Collection, Home Insurance Company, New York

Company No. 34 known as 'Red Rovers' from Greenwich Village, is operating a New York-style goose-neck, side-stroke engine, while on the right, Engine Company No. 14 from Vesey Street appear to have temporarily ceased working their Philadelphia-style, double-decker, end-stroke engine.

Several foremen with raised arms orchestrate, like conductors, the labours of their teams. The foremen's shouted instructions – 'Stand by your brakes ... ... play away men ... work her lively lads' – are recorded in crude hand-printed lettering in margins around the image. From this we learn that firemen ascribed a female gender to their engines. Robyn Cooper has remarked: 'Like ships, trains and other machinery, the fire engine is feminised because of its "beauty" and subjection to masculine power and control. Moreover, these feminine machines were always obedient and obliging unlike the women the firemen were wedded to.'

The statement 'old 14 washed by 34' describes how Engine Company No. 34 defeated its rival No. 14. 'Washing' refers to a flooding of an engine by another to which it is linked by hoses (in order to relay water from a distant source) caused by one engine crew working its brakes harder and faster than the other crew. To be washed was considered a public humiliation and beaten fire laddies were known to weep or to start a fight when it happened. The image's strange title – 'Up to the bend' – refers to the depth of water in the well or feeder-box of a fire engine. The late Harold V. Smith collected these two images when he was President of the Home Insurance Company from 1937 to 1954. Smith established a museum of firefighting memorabilia, which was housed on the 15th floor of the Company's headquarters at 59 Maiden Lane, New York. In 1980, the Company disposed of its collection, but much of it was donated to the New York City Fire Museum.

30 Gustave Courbet, *The Departure of the Fire Brigade*, (1850–51). Painting, oil on canvas, 388 x 580 cms. Paris: Musée du Petit Palais. Photo courtesy of the Photothèque des Musées de la Ville de Paris

## 30 — 1850–51

### Gustave Courbet, The Departure of the Fire Brigade

During 1850–51, the French artist Gustave Courbet (1819–77) worked on a huge canvas entitled *The Departure of the Fire Brigade*. Set in the crowded streets of Paris, it depicts a murky evening scene and shows what appear to be three teams of firemen equipped with manually-operated pumps. Since the men are pulling pumps by hand, it would seem the Paris brigades were not yet equipped with horse-drawn appliances. (The date is too early for steam pumps and so the flames in the background must be from braziers.) Since the teams appear unsure what direction to take, a fire officer and a civilian on the right of the picture point the way. In the foreground is a mysterious crouching figure holding one hand up towards the viewer. He is in imminent danger of being run over by the advancing firemen who are flanked by representatives of two classes in French society: on the extreme left, there is a poor working-class mother and child, and on the extreme right there is an affluent bourgeois couple. In terms of colour, the painting is organised around a contrast between yellow and dark blue.

Courbet scholars maintain that the painting was partly inspired by the artist observing firemen in action putting out a fire at a police station set alight by revolutionaries during the 1848 uprising that led to the Second Republic, and partly by Rembrandt's famous canvas *The Night Watch* (1642). Both paintings depict urban night scenes involving people engaged in civic duties.

Courbet had the approval of the War Ministry and the cooperation of Commander Victor Frond who served as the model for the fire chief. Frond was in charge of a fire station in Rue Saint-Victor and arranged for Courbet to witness an alarm demonstration for authenticity's sake. Courbet was one of many artists and intellectuals inspired by new ideologies such as anarchism and socialism who became fascinated by folk culture, and who celebrated 'the people', that is, peasants and workers. Admirers promoted Courbet as a 'socialist painter' and he styled himself a 'realist'. At the time, there was also an anthropological-type urge to document and categorise trades and professions, which naturally included firemen. It was during the nineteenth century that fire services in major cities ceased to be primarily staffed by volunteers and became fully professionalised and mechanised. The moment of urgency Courbet selected, when brigades emerged at speed from fire stations with engines, soon became a favourite subject for visual artists, illustrators and photographers in both Europe and the United States.

Despite Courbet's commitment to observing and recording the surface appearances of reality, he also liked to comment on leading figures and political events in France by including an allegorical dimension in some of his largest pictures. (It was an indirect method of communication he adopted to circumvent censorship; its drawback was obscurity of meaning.) Comprehending the political aspects of *The Departure* ... is now problematic. One reason it is hard to decipher is that it was never finished. Significantly, Courbet abandoned the painting because of the arrest of Frond for encouraging his men to resist Louis Napoleon's *coup d'état* of December 1851, which ended a republican regime. Courbet, a keen democrat and republican, no doubt feared his association with Frond and the previous regime would place him in danger if he continued with such a populist painting.

## 31 — 1855

## Sir John Everett Millais, The Rescue

In Courbet's canvas, the collective action of a brigade is depicted, whereas when the English Pre-Raphaelite painter Sir John Everett Millais (1829–96) tackled the subject of firefighting in 1855, he preferred to focus on individual heroism. His dramatic painting *The Rescue* shows a square-jawed London fireman descending a carpeted staircase inside a house to escape a blaze behind him. Fire has violated a Victorian family's domestic tranquillity. A middle-class home is indicated by the carpet's width and the richness of its design, and by the elegance of the mother's dress. The fireman is dressed in a plain, functional, dark uniform with black leather helmet and boots introduced by James Braidwood, Superintendent of the London Fire Engine Establishment. (The latter organisation was founded in 1833 from a number of insurance brigades. Although still funded by the insurance companies, it marked the beginning of paid, full-time, professional firemen serving the whole community.) The fireman is carrying no less than three children – a baby under his left arm, a girl under his right and a boy clinging to his back – towards the outstretched arms of their mother whose expression is a mixture of anxiety and relief. (Mysteriously absent is the father.) Millais is surely contending that what matters now to the fire service, more than preserving property, is saving human life. The left hand side of the canvas is warm/masculine, while the right hand side is cool/feminine. Behind the mother, a cold sky and rooftops can be seen though an arched window. Furthermore, the oblong canvas was itself mounted in a frame arched at the top to evoke an architectural setting.

According to his brother, Millais was inspired to paint *The Rescue* after witnessing the deaths of two firemen during a blaze in London at Meux's Brewery in Tottenham Court Road one night during January 1854. Furthermore, he thought that while soldiers and sailors had received plenty of attention from artists, firemen had not. (At the time, British soldiers were abroad fighting in the Crimean War. Firemen resembled soldiers because they were uniformed, disciplined bodies of men who performed heroic deeds, but instead of taking lives like soldiers sometimes do, they saved them.)

Millais first produced some sketches that are now in the Courtauld Institute Galleries and then used a pencil and red chalk to produce a cartoon of the composition, which is now in the collection of the Birmingham Museum & Art Gallery. The painting itself took three months to execute and was Millais' first completed work tackling a contemporary subject (some Pre-Raphaelites looked back to the Bible or the Middle Ages for subject

31 Sir John Everett Millais, *The Rescue*, (1855). Painting, oil on canvas, 121.5 x 83.6 cms. Melbourne, Australia: National Gallery of Victoria, Felton Bequest 1924. Photo courtesy of the Gallery

matter), which he researched by accompanying Braidwood to fires. He also consulted the novelist Charles Dickens regarding the subject matter.

Obviously, the canvas was not painted from a directly observed event. It was an imagined scene Millais recreated in his studio using models. To capture the light and smoke effects of fires, he used a sheet of coloured glass and burnt planks on an iron sheet placed on the floor. Millais' friend and fellow artist Charles Collins painted the hosepipe visible in the painting's bottom left-hand corner.

Millais hurried to finish the painting by April 1855 so that he could submit it to a Royal Academy exhibition. Later, the canvas was exhibited in Glasgow, Liverpool and Paris. According to his son, Millais considered *The Rescue* his best picture. William M. Rossetti and John Ruskin praised it and one critic described the fireman as 'a hero of this nineteenth century – a soldier of the fire brigade'. However, some viewers thought that the fireman's head was too big for his stumpy body and that the figure of the mother was insubstantial. A lawyer named Joseph Arden bought the painting for £580. Surprisingly, no print of it seems to have been produced and sold during the late 1850s although a wood engraving did appear in volume one of the *National Magazine* in 1857. In 1882, the Liverpool Shipwreck & Humane Society, a charity founded in 1839, used Millais' painting as the basis of a design on Fire Medals, cast from gold, silver and bronze, awarded to worthy firemen.

*The Rescue* is now in the collection of the National Gallery of Victoria, Melbourne, Australia. In 1986, Robyn Cooper, of Sydney University, wrote a perceptive analysis of the painting in which she argued that it was 'a site for the interaction of complex and conflicting discourses – about fire, home, the family, sex, gender and class – which could not be fully resolved into a single, harmonious and unproblematic statement'. She also maintained that Braidwood's men were still concerned with preserving property rather than human life and that the latter responsibility belonged to members of the Fire Escape Society formed in London in 1828. The Royal Society for the Protection of Life from Fire (RSPLF), founded in 1836, later absorbed it. The RSPLF established 85 fire posts equipped with 'escapes' (ladders, canvas chutes, etc.) operated by men who were called 'Conductors'.

After 1855, the 'rescue' or 'saved' motif was to become one of the most common in firefighting iconography. In the main, babies, girls and young women were saved rather than boys or men, but sometimes the victim was an animal such as a cat.

32 Pierre Puvis de Chavannes, *The Village Firemen*, (1857). Painting, oil on canvas, 179 x 228 cms. St Petersburg: Hermitage Museum. Photo courtesy of the Museum

## 32 1857

## Pierre Puvis de Chavannes, The Village Firemen

In 1857, the French artist Pierre Puvis de Chavannes (1824–98) painted *The Village Firemen*, which in contrast to the urban firefighting scenes of Courbet and Millais tackled a rural subject. A drawing in red chalk and charcoal also exists, which served as a study for the painting.

Puvis is best known for his pale-coloured allegorical murals featuring statuesque figures that decorate public buildings across France. He was more of an idealist than the realist Courbet because he generalised his figures and worked from memory and his imagination. However, the village firemen in this painting, who were almost certainly part-time volunteers recruited from local artisans, are not statuesque because they are dragging a heavy manual pump across broken ground towards burning thatched cottages on the horizon which are generating much smoke. The weather appears stormy with wind and rain. The firemen have managed to obtain metal fire helmets from the Ministry of Interior but not uniforms because they are still wearing peasant smocks. While most figures head towards the fire, others flee laden with goods. In the bottom left-hand corner, two women help to push the pump out a gully in which it has become bogged down. Those hauling the pump are followed by a crowd, which includes more women, carrying such firefighting equipment as buckets and ladders. A soldier and a priest also appear. Puvis is surely reflecting the fact that rural *corps de sapeurs-pompiers* were manifestations of civic responsibility, co-operation and fraternity, and that fighting fires often required the mobilisation of the whole community.

33 Unknown sculptor, *Harry Howard*. (c. 1857). Polychrome painted and carved wood, 254 x 168 x 112 cms; accession #1937.328. Collection of The New-York Historical Society

Historians claim that rural fires were frequent major hazards during the nineteenth century, that many were started deliberately for political, malicious or financial reasons, and that Puvis was inspired to address the subject after observing such a fire from a train window while passing through the Mâconnais region. While increasing numbers of commune councils established firefighting companies and bought pumps for them as the nineteenth century progressed, there were many communes that either could not afford them or could not find enough volunteers; consequently, one commune's company was expected to help its neighbours. Clearly, there were often long delays before such companies could reach fires in scattered hamlets and farms, with the result that in 1857 some conflagrations destroyed whole villages. Puvis' composition resembles war paintings of the 1790s in which troops drag cannons and anticipates Soviet socialist realist pictures in which gangs of labourers haul barges up rivers. Appropriately therefore, the painting is in the collection of the Hermitage Museum, St Petersburg.

## 33 — 1850–60

**Unknown sculptor, Wooden Fire Chief**

## 34 — 1854

**Endicott & Co, Fireman's Hall, New York**

## 35 — 1857

**Joseph Hoffman Johnson, Harry Howard, Chief Engineer**

This statue of a famous and highly respected American fire chief made from carved and painted wood stands on a square pedestal and is part of the collection of the New York Historical Society. The name of the sculptor is not known but the man depicted is thought to be Harry

34 Endicott & Co (lithographers and printers, New York), *Fireman's Hall*, New York, erected in 1854. Print/bookplate, with figure of fireman crowning the Hall's frontage

Howard (1822–96), Chief Engineer of New York's Volunteer Fire Department from 1857 to 1860. (A 'chief engineer' was an overseer of fire engines.) Howard, wearing a shirt which still retains some reddish colouring, a fire hat and a knee-length frock coat, is shown in dynamic pose with both arms raised; one is pointing and the other holds a speaking trumpet or horn up to his mouth. (As already explained, prints of fire scenes in the nineteenth century often included a foreman in a comparable pose.) Evidently, he is directing operations.

The statue is thought to have originally stood in front of the Fireman's Hall at 127–29 Mercer Street, New York. Nineteenth-century prints reveal that when a new Hall was erected at 155 Mercer Street in 1854, the statue was placed on the top of the building's façade. (But shown in reverse due to the printing process?) However, another illustration depicting the last Grand Triennial Parade of fire companies held on 17 October 1859 shows a similar statue in place in front of Howard's New York home.

Since 9/11, the New York Historical Society has paid homage to fallen firefighters by mounting memorial exhibits in which the wooden statue of Howard has been displayed in front of a huge American flag. Immediately following 9/11, the NYHS began to collect objects, images and ephemera associated with the event

35 Joseph Hoffman Johnson, *Harry Howard, Chief Engineer*, (1857?). Painting, oil painting on canvas. New York: The New York City Fire Museum. Photo courtesy of the Museum

for the benefit of future generations.

A full-length portrait by Joseph Hoffman Johnson of a clean-shaven Howard also exists that shows him in a similar pose but wearing a double-breasted, tan-coloured 'turnout' coat. The artist has clearly delineated his subject's handsome face and determined expression. Howard stands in a large open space and his figure dominates the painting's foreground. Behind him, a building is burning furiously and emitting a huge column of dark smoke. Fire companies are tackling the blaze. Apparently, the incident depicted was a fire at the William T. Jennings & Co. clothing store at 231 Broadway, which occurred during April 1854 when Howard was in charge of Engine Co. No. 21. Tragically, 13 firemen were killed when the building suddenly collapsed.

Another portrait of an older Howard, this time sporting a handlebar moustache and wearing a fire hat shield labelled 'Chief Engineer', again painted by Johnson, is preserved in the American Museum of Firefighting, Hudson River Valley. An ambrotype photograph (a collidion process patented in the United States in 1854) of Howard, a full-length portrait taken by famous American photographer Mathew Brady, survives. Furthermore, Howard appears in group portraits depicting New York's fire chiefs. Painters and photographers probably portrayed Howard more frequently than any other American fire chief.

Howard the man, according to one of his contemporaries, had 'a Herculean frame, an iron constitution, and nerves of steel', plus a kind and generous disposition. His interest in fires began in childhood: at the age of 13, in 1835, he witnessed the Great Fire of New York in the financial district of Wall Street and helped as a runner. (Runners called out volunteer firemen from their homes or workplaces and helped to pull ropes attached to fire engines.) In 1841, he became a fully-fledged member of the Petersen Engine Company No. 15. Gradually, he rose through the ranks until, in 1857, he was elected Chief Engineer, a post similar to that of a modern Fire Commissioner. He was responsible for 4000 men and was instrumental in introducing bunkrooms in the city's firehouses. Bunks enabled firemen to sleep in firehouses overnight and so they could respond to alarms more quickly. As a result, more buildings and lives were saved and insurance rates fell. Howard advocated paying firemen for their services and reducing their workloads by using steam pumps and by employing horses to pull heavy equipment.

Unfortunately, in 1858, he was disabled after suffering a stroke while hurrying to a fire. Later in life, he helped to found the Firemen's Home of the Firemen's Association of the State of New York, a charitable organisation dating from 1884. His funeral in 1896 was one of largest gatherings of firemen, friends and officials New York has ever seen. Howard is buried across the

36 Joseph Hoffman Johnson, *Portrait of John Baulch*, (1850s?). Painting, oil on canvas, 72 x 91 cms. Hudson, New York: American Museum of Firefighting. Photo courtesy of the Museum

road from the Fireman's Monument in Green-Wood Cemetery, Brooklyn and his memory lives on in the names of fire companies, squares and avenues.

## 36 — 1850s

### Joseph Hoffman Johnson, Portrait of, John Baulch

## 37 — 1886–87

### Joseph Hoffman Johnson, Banner for the Volunteer Firemen's Association in New York

Johnson, the creator of this half-length portrait of a leading American fireman, not only painted portraits but also picture panels and decorations on fire engines and carriage reels. In addition, he painted a large banner, which was presented to the Fire Department by the City of New York in 1858, which featured the allegorical female figures Protection and Benevolence surrounded by emblems of the Fire Department. Another painted banner, five-by-seven feet made from heavy double blue and scarlet silk, which Johnson executed in 1886–87 for the Volunteer Firemen's Association in New York (incorporated 1884), employed the same allegorical figures plus a stalwart fireman holding the hands of two orphan boys.

The portrait's subject was John Baulch, a robust individual with impressive sideburns who was a long-serving fireman. He occupied positions of assistant foreman and foreman, assistant and senior engineer, with several companies and replaced Chief Engineer Harry Howard for a time when the latter became ill. Baulch is said to have attended every large fire in New York between 1835 and 1862.

Volunteer firemen were normally exempt from military service but during the American Civil War, many joined up. Baulch organised fire companies at Fortress Monroe, Hampton, Virginia. His task was to follow the Union Army and take charge of fire equipment in captured towns and to prevent any needless destruction of property. Two silver trumpets were presented to him by fellow firefighters in New York and Philadelphia and he eventually became vice-president of the Association of Fire Engineers of the United States.

In Johnson's rather crude portrait, executed during the 1850s, Baulch is shown wearing the red tunic and helmet of Manhattan's Engine Company No. 13, which has his initials and rank on it. He is also holding, in his

37 Joseph Hoffman Johnson, *Banner for the Volunteer Firemen's Association in New York*, (1886–87). Engraving of front side of painted silk banner, 152.5 x 213.5 cms

38 Unknown photographer, *Portrait of a Member of the Vigilant Fire Company, Baltimore*, (between 1840–60). Hand-coloured, sixth-plate daguerreotype. Washington, DC: Collection of the Library of Congress, Prints and Photographs Division, Marian S. Carson Collection

right hand, a speaking trumpet as a sign of his position as foreman; in his left, there is a half-smoked cigar. Rather unconvincingly, the portrait is set out of doors. On the left, there are some trees and on the right, a row of buildings silhouetted against the light of a fire. The painting of this courageous, humane and popular fireman is now preserved in the American Museum of Firefighting, Hudson, New York.

## 38 1840–60

## Unknown photographer, Portrait of a member of the Vigilant Fire Company, Baltimore

## 39 1855–56

## Unknown photographer, Fireman in uniform holding a brass instrument

The invention of photography during the first half of the nineteenth century greatly democratised and expanded the genre of portraiture. One type of photographic process, a direct-positive process known as the daguerreotype, was announced in 1839 and became popular in Europe and America during the period 1840–70. It was named after its inventor, the French painter Louis J.M. Daguerre (1787–1851). Each daguerreotype was unique because there was no negative to facilitate replication. Daguerreotype images were formed on copper plates coated with silver iodine and so were subject to surface scratching. Their reflective surfaces gave them a mirror-like quality. Many that survive have pale colours because they were tinted by hand. To protect them and present them to best advantage, they were covered with glass and enclosed in decorative leather or paper frames or portable thermoplastic cases. Some photo cases moulded from gutta-percha (the inventor of thermoplastic cases in 1854 was Samuel Peck) featured images of firemen and fire engines in relief. Other kinds of early photographs were called ambrotypes, tintypes and albumen. The latter were generated from wet plate negatives and so copies of the same image could be made and mounted on paperboards.

Commercial photographic studios were quickly established and many people visited them to have their photographs taken dressed in the attire, and with the tools, of their particular trades or professions and this included firefighting. Among the ones that survive are portraits of individual firemen (half- or full-length), pairs and threesomes. In a few instances, whole companies were photographed out of doors while forming a semi-circle around their fire engines.

As indicated by the letter 'V' on his chest and the word 'vigilant' on his painted, stovepipe, parade hat and belt buckle, the anonymous young man depicted in this frontal view, unattributed daguerreotype was a member of the Vigilant Fire Company of Baltimore. The motif of the eye on the hat stemmed from the Company's motto 'The Eye Of Jackson was for his country's glory Ever Vigilant'. The fireman was probably a foreman because of the speaking trumpet in his right hand. In his belt there is a spanner. Other props found in portraits are: axes, brass nozzles, capes, ornate uniforms, hats, wooden fire hydrants and even model fire engines.

An unusual, hand-coloured daguerreotype owned by the John Paul Getty Museum in California depicts an African-American fireman who is holding a brass musical instrument – a cornet – presumably for use in a marching band accompanying the public parades fire companies were so fond of holding. The man also wears a ribbon with a medallion around his neck; consequently, he has won an award of some kind.

Most early portraits were taken in studios equipped with head clamps and required long exposure times, and so poses and facial expressions tended to be rather solemn and stiff. Furthermore, uniformed firemen posing among the standard studio props and backdrops, such as pedestals and small tables placed in front of a painted landscape with classical buildings, yielded incongruous juxtapositions.

Today, however, such portraits greatly appeal to American collectors, one of the best known of whom is Michael J. Novak, who for 20 years was an engineer for the Newport Beach Fire Department, California. He is also the author of scholarly articles on daguerreotype fireman portraits from the era when companies consisted of volunteers.

39 Unknown photographer, *Fireman in uniform holding a brass instrument*, (c. 1855–56). Hand-coloured daguerreotype, 8 x 7 cms. Photo © 2003 The J. Paul Gety Museum, Malibu, California.

40 John Hubert Newman (?), *Portrait of Andrew Torning*, (1850s or 1860s). Dry plate photograph, 31 x 25.5 cms.
Melbourne, Australia: State Library of New South Wales © 2003

## 40 — 1850s or 1860s

### John Hubert Newman (?), Portrait of Andrew Torning

The subject of this oval-shaped, half-length portrait was an early Australian fire chief called Andrew Torning. A well-travelled man, he was born in England in 1814 and died in Australia in 1900. He arrived in Sydney with his actress wife Eliza and a troupe of actors in 1842 and remained there until 1858. Then he journeyed to San Francisco where he and his family lived for almost a decade. Torning was a versatile individual: a painter of townscapes, a set designer, an actor and a keen volunteer fireman. In the portrait, his calm gaze, assured pose and evident pleasure in wearing a fireman's uniform replete with metal buttons and a medal may be attributed to his thespian profession.

As an actor, Torning must have been conscious that many Victorian theatres were firetraps and this was probably why he joined the fire brigade of the Victoria Theatre, Sydney in the 1840s. In 1854, Torning took over the lease of the 'Vic' and formally constituted the Victoria Volunteer Fire Company No. 1, whose motto was 'We strive to save', which was soon renamed the Australian Volunteer Fire Company No. 1. Another photograph by Newman shows Torning standing proprietarily alongside a Pioneer manual pump he imported in 1855 at a cost of £500. In 1857, Torning was also instrumental in the building of Australia's first, purpose-built, brick fire station in Pitt Street South, Sydney.

While in the United States, he maintained his enthusiasm for firefighting. In fact, he became an American citizen so that he could join the San Francisco Fire Department. On his return to Sydney, he was re-elected Superintendent of the Australian Fire Company No. 1. This explains the presence of the brass speaking trumpet and the number one on his helmet and belt. Torning was notorious for shouting orders to his men via the trumpet, which is now in the collection of Penrith's Museum of Fire. During the 1870s and 1880s, Torning became embroiled in the politics concerning the future of the fire service. He favoured independent volunteer brigades serving the whole community rather than profit-orientated brigades paid for by private insurance companies. Since his death, Torning has been hailed as 'a folk hero', 'humanitarian' and as 'the founder of the volunteer fire brigades movement in Australia'.

41 Stett & Martin (engravers), *Rob Tempest & Jos Barton, President and Vice President of the Hibernia Engine Co.*, (1858?). Engraved, coloured and printed for the *Firemen's Magazine*

## 41 — 1858

### Stett & Martin, Rob Tempest & Jos Barton, President and Vice President of the Hibernia Engine Co.

## 42 — 1851

### James Fuller Queen, Weccacoe Engine Cos. House and Reception of United Fire Co. of Baltimore

The first print depicts the colourful and picturesque uniform adopted by the officers of some early American volunteer fire companies, in this instance, the Hibernia Fire Engine Company No. 1 of Philadelphia, which was founded in 1752. The term 'Hibernia' is Latin for Ireland and so most members of the Company were Irish-Americans. Presumably, the fondness for the colour green was also a tribute to Ireland. Barton, depicted in profile, holds a brass speaking trumpet while Tempest, depicted from the front, holds a book labelled 'Constitution' (that is, the Constitution of the Company), which stated that each member must provide two leather buckets, two salvage bags and a large wicker basket. The officers' stovepipe hats were painted

42 James Fuller Queen (draughtsman), *Weccacoe Engine Cos. House and Reception of United Fire Co. of Baltimore*, (1851). Lithograph, 57.8 x 40 cms. Printed and published by P.S. Duval Steam Lithographic Press, Philadelphia. Photo: courtesy of The Maryland Historical Society, Baltimore, Maryland

with the Company's emblem – an eagle grasping a harp – as were the capes intended to protect them from water. Such capes are thought to date from the late eighteenth century. The letters 'F' and 'A' probably stand for 'Fire Association'. Both officers are rather grim-faced and stand stiffly like mannequins modelling new clothes in a fashion plate. Their attire does not appear to be suitable for firefighting and hence probably represents their ceremonial or parade costumes.

Hibernia occupied a four-storey firehouse with a neoclassical façade and owned a double-decker, end-stroke manual pump made by John Agnew of Philadelphia during the 1830s. The engine was richly decorated with panel paintings and there is a lithograph by James Fuller Queen (1820–86) dated 1858 showing the Company proudly standing next to it outside the firehouse. The Company acquired its first steam engine in 1858 and, in the same year, its 127 men visited Brooklyn, Charlestown, Newark and Boston where demonstrations, parades and speeches were made. To commemorate this trip, J.B. Chandler of Philadelphia printed an illustrated volume entitled *The Hibernia Fire Engine Company No. 1* (1859).

Queen, the artist, was a native of Philadelphia noted for his watercolours and lithographs, portraits, landscapes, advertisements, certificates and images of war and disaster. In 1835, at the age of 15, he was apprenticed to the firm of Peter S. Duval & Son. Duval (1804 or 5–86) was born in France, trained as lithographer in Paris and came to the United States around 1830. His steam-powered lithographic press of Philadelphia was one of the first in the United States to print colour lithographs. Firefighting was a subject Queen had previously addressed at least three times: in 1851 he produced a lithograph showing members of the Weccacoe Engine Company of Philadelphia in front of their three-storey firehouse welcoming visitors from the United Fire Company of Baltimore; and in 1857 he painted one watercolour depicting firemen departing from a firehouse to answer a night alarm and another showing a company pulling its engine through a moonlit street.

## 43 1858

## Louis Maurer, The American Fireman. Prompt to the Rescue

Following Millais' *The Rescue*, artists in Europe and North America repeatedly depicted firemen emerging from flaming rooms/houses or descending steps/ladders carrying children or maidens in their arms (elderly men and women seem not to have been pictured). Females were usually shown unconscious with tresses of hair hanging down and clad only in nightdresses. Examples include, Louis Maurer's *Prompt to the Rescue* (1858), a Currier & Ives lithograph from 'The American Fireman' series depicting a red-shirted fireman with a girl cradled in both arms fleeing from a bedroom that is aflame; Thomas Waterman Wood's painting *The Rescue* (1870); and Charles Vigor's painting *Saved* (c. 1890).

Nathaniel Currier (1813–88), a lithographer, established a print business in New York in 1834. He employed fine artists to generate sketches, which lithographers transferred to stones; hand-pulled prints were then coloured by female employees. Editions were 'unlimited' with a minimum of one hundred. Currier had a personal interest in the subject of firefighting because as a young man he had been a volunteer member of the Excelsior Company No. 2: he fought the Great Fire in New York's financial district in December 1835 and soon afterwards issued a print entitled *The Burning of the Merchants' Exchange*. The firm Currier & Ives was formed in 1857 when the self-taught artist James Merritt Ives (1824–95) became Currier's partner. It quickly became famous for its many series of lithographs depicting aspects of American life and it survived until 1907. In fact, Currier & Ives have since been dubbed 'Printmakers to the American People'.

43 Louis Maurer, *The American Fireman. Prompt to the Rescue*, (1858). Hand-coloured lithograph published by Currier & Ives, New York, 44.1 x 34.6 cms. Library of Congress, Prints and Photographs division

The four-print series 'The Life of a Fireman' (1854, plus two further prints dated 1861 and 1866 by Charles Parsons and John Cameron; reprinted in 1884) and the six-print series 'The American Fireman' (1858) were especially popular and served to disseminate the fire-man-as-hero image throughout the United States and Europe. Currier was himself depicted as a fireman pulling an engine in Maurer's *The American Fireman. Always Ready* (1858). Some of Currier & Ives' prints focus on individual firemen, but others show companies leaving firehouses, dragging their engines by hand through the streets or using horses to pull steam engines and trucks, and tackling blazes with pumps, hoses and ladders.

Maurer (1832–1932), the son of a cabinetmaker, was born in Germany and in childhood developed a love of drawing. He came to the United States with his family in 1851. At first, he worked as a woodcarver but then switched to lithography. Currier & Ives employed him for eight years (1852–60) before he founded his own firm. Maurer was especially skilled at depicting horses and eventually became a painter of genre scenes.

Maurer's 'saved' composition, which of course occurs in reality when rescuers carry injured or dead bodies (witness the photograph of firefighter Chris Fields holding a dead baby taken in 1995 by Charles Porter, IV in Oklahoma City), was to prove a powerful icon, a precedent for many subsequent images and sculptures.

A curious homage to 'The Life of a Fireman' prints occurred in 1971 when the Spanish surrealist Salvador Dalí (1904–89) produced six interpretations of them in the medium of gouache, which formed the basis for lithographs.

The Phyllis Lucas Gallery of New York published the latter in a limited edition of 305. A frequently reproduced example – *The Life of a Fireman. Fire! Fire! Fire!* – includes (as a collage) the Currier & Ives' print subtitled *The Fire. "Now then with a will – shake her up boys"* and surrounds it with further windows bursting with flames, more firemen with ladders plus a large death's head sphinx moth (a symbol of mortality) which is attracted to, but may survive, the flames. The original gouache was executed in an expressionist rather than a surrealist manner. Aside from commercial expediency, the point of such repetition and appropriation of past popular culture is hard to fathom.

## 44 — 1861

### Unknown graphic artist, Memorial to James Braidwood

## 45 — 1861

### Unknown artisan, Braidwood

By the 1830s, it had become clear that to fight fires in London more effectively, the various private insurance company brigades needed to be unified into a single organisation and that it would need a chief executive. Thus, the London Fire Engine Establishment (LFEE) was founded in 1833 and the post of Superintendent created. Because of their wide responsibilities and the importance of their post, London's fire chiefs during the nineteenth century were public figures whose characters, lives and deeds were much commented upon in the print media of the day. Cartoonists, ceramicists, illustrators, painters, photographers and sculptors portrayed them.

James Braidwood (1800–61), a Scotsman who had been Edinburgh's Firemaster for nearly a decade, and who was a qualified surveyor and engineer, was appointed as the LFEE's first Superintendent. He took charge of 80 full-time firemen operating from 13 stations. Portrait photographs show a handsome man with a serious but kindly face, long nose and prominent sideburns. Braidwood proved to be a conscientious manager, an experimenter in firefighting techniques (although he was slow to appreciate the value of steam-powered fire engines), and an educationalist (he wrote a manual on fire engine construction and training methods). His men – soon known as 'Jim Braidys'– were encouraged to train hard, to practice at night and to enter buildings to find and extinguish the sources of fires rather than directing jets of water from a distance (a method called 'surround and drown' in the United States). Braidwood became so well known that his name was mentioned in music-hall songs and the public cheered him when he rode engines with his men through London's streets.

44 Unknown graphic artist, *Memorial to James Braidwood*, (1861). Hand-coloured print, 21 x 16 cms. Published by *The British Workman*

45 Unknown artisan, *Braidwood*, (c. 1861). Staffordshire ceramic figure, 38 cms high. Stoke-on-Trent: The Potteries Museum & Art Gallery. Photo courtesy of the Museum

*The British Workman*, a journal published from 1856 to 1921, issued the anonymous coloured engraving reproduced. It is a pictorial memorial to Braidwood because on 22 June 1861 he was killed by falling masonry while leading his men trying to contain a massive fire at Cotton's Wharf, Tooley Street on the South Bank of the Thames. A view of the inferno – the worst since the

Great Fire of London in 1666 – from across the river appears beneath his portrait (several prints by different artists exist of this scene). The location of the fire was an area of warehouses, separated by narrow alleyways, which were full of inflammable goods and so the conflagration was immense. It spread to other wharves and even ships moored on the river. Smoke and flames could be seen for miles and so the fire, which raged for days, attracted thousands of sightseers.

The portrait of Braidwood shows him wearing the black leather helmet and dark uniform, which he introduced. The shoulder epaulettes were found under rubble where his body had lain for three days and are now preserved in the London Fire Museum. He is flanked on either side by allegorical figures representing Fame (on the left) who wears a sash stating 'Honour to the brave' and Britannia (on the right) who wears a sash stating 'A nation mourns him'. Even Queen Victoria sent her condolences to his family. Thousands of mourners watched Braidwood's funeral procession, which was a mile and a half long. A lithographic print depicting the cortege as it passed through crowds near the Royal Exchange was created by artist A.L. Butler and published by C.J. Culliford (Guildhall Library Print Room has one). Braidwood was buried in the Abney Park Cemetery, Stoke Newington, in north London.

46 Unknown photographer, *Portrait of Lillie Hitchcock-Coit*, (c. 1863). San Francisco: California Historical Society. Photo courtesy of CHS

The death on duty of such a prominent fire officer was indeed a shock to the nation. The following year, a marble memorial tablet to Braidwood was placed on a wall in Cotton Street, near where he died. Samuel Henry Gardiner – who seems to have been a local stonemason based in New Kent Road – carved the sculpture, which was commissioned by the Metropolitan Police. It features an inscribed dedication enclosed by a laurel wreath behind which are buildings belching flames and smoke. At the base of the wreath, firefighting equipment appears: a helmet, a trumpet, an axe and a rolled hose. Staffordshire ceramic figures of Braidwood were also produced and London streets and fireboats were named after him. (Even today, a model-kit of a 1930s' James Braidwood London Fire Boat can be purchased from Caldercraft.) It is no surprise, therefore, that a biography of Braidwood by Brian Henham published in 2000 was entitled *True Hero*.

## 46 1863

**Unknown photographer, Portrait of Lillie Hitchcock-Coit**

## 47 1860

**Stephen C. Massett, No. 5's Gal [caricature of Lillie Coit]**

## 48 1940s

**Stanley A. Piltz Company, Coit Memorial, Telegraph Hill, San Francisco,**

This may be the earliest photograph of a female firefighter or 'firebelle', albeit an honorary one. Taken by an unknown photographer around 1863, it depicts Elizabeth Wyche 'Lillie' Hitchcock-Coit (1843–1929) wearing a fire hat (which has her initials LHC on it), a badge, a shirt and belt buckle of the Veteran Firemen's Association of California. Posed rather awkwardly on a wooden chair set against a plain background, she gazes to the right and there is a hint of a smile on her lips. The impression conveyed is of a confidant, capable and physically active woman. One can perhaps guess that in her youth she was a high-spirited tomboy fond of pranks and escapades; someone prepared to break social conventions particularly those applying to gentlewomen.

Female firefighters in the United States predate this photograph: for instance, in 1818 a black slave called Molly Williams assisted the Oceanus Engine Company No. 11 in New York and, during the 1820s, Marina Betts, a woman of French-Indian origins, was a keen volunteer firefighter in Pittsburgh. Betts worked on

47 Stephen C. Massett, *No. 5's Gal* [caricature of Lillie Coit], (1860). Pen and ink drawing. Oakland, California: Hitchcock-Coit Papers Collection, F.W. Olin Library, Mills College

bucket lines and fought fires for ten years. She became notorious for dousing male bystanders with water when they refused to help.

Hitchcock-Coit, a colourful and eccentric woman from a Southern upper-class background, came to San Francisco aged seven in 1851 from Maryland. The Hitchcock family arrived soon after a major fire had devastated what was then a wild, gold rush, frontier town. (San Francisco, of course, is a city with a long history of disastrous fires, especially in 1906, aggravated by high winds, and by earthquakes that disrupt water supplies.) Shortly afterwards she was playing with friends in a half-built hotel when it caught alight. Two of her playmates died but John Boynton from the volunteer Knickerbocker Engine Company No. 5 (founded in 1850) saved her. Firemen – whom she once described as 'courage in uniform' – immediately became her heroes. Later, as a teenager, she witnessed men from the same Company struggling to pull its engines up Telegraph Hill (they were short-handed at the time). She called for help from male bystanders and set an example by grabbing the tow-rope. As a result, she became the Company's mascot.

A roughly drawn caricature survives of Lillie running with her pigtails flying. She grasps a rope in her right hand and with the other holds a speaking horn to her lips. She shouts: 'This way boys! Follow me!! I'm yer gal!!!' A caption reads: 'Fastest gal of the age – time 6.20 to a hose cart'. The draughtsman was Stephen C. Massett (1820–98), aka 'Jeems Pipes of Pipesville', an English-born actor, author and composer who lived in California for a time. Massett, who was a member of Engine Company No. 5, presented the caricature to 17–year-old Lillie on 14 February 1860.

In 1863, the firemen presented her with a certificate of honorary membership of the Company and a gold, diamond-studded badge reading 'No. 5', which she wore on her red fireman's shirt. She attended many of the fires the Company tackled (to cheer the men on), the gala parades in which it participated and the banquets it organised. Lillie greatly enjoyed the company of men and had many admirers (at that time, the ratio of men to women in San Francisco was around 50–to–1). She was fond of riding and driving fast horses, staying up all night smoking cigars and playing poker with male friends but she also assisted injured firemen and helped the families of the fallen.

During 1862–63, while the American Civil War was

48 Stanley A. Piltz Company, *Coit Memorial, Telegraph Hill, San Francisco*, (1940s?). Linen, colour, picture postcard, 9 x 14 cms. No. 29, 'Pictorial Wonderland', Art-Tone Series. © Stanley A. Piltz, San Francisco. Author's collection

49 Unknown photographer, *Hand-in-Hand Fire Company No. 1, Philadelphia, with their Steamer*, (c. 1863). Philadelphia: CIGNA Museum & Art Collection

raging, Lillie – who supported the South – spent time in Paris at the court of Emperor Napoleon III. One of her first acts on arrival in Paris was to visit a fire scene! Invited to a masked ball, she devised a firefighter costume, which she later wore in San Francisco. (It may be the one she is wearing in the portrait.) On another trip to Paris in 1865, she visited the studio of the noted woman painter of horses, Rosa Bonheur. In 1868 Lillie married Howard Coit, a financier, and when he died prematurely in 1885 at the age of 47 she became a wealthy widow who spent decades travelling the globe and living in France.

A portrait of Lillie wearing more conventional female attire painted in the 1870s is preserved in the San Francisco Fire Museum. However, even this portrait includes an enamel brooch with 'No. 5' on it. In the 1890s, she also presented a bust of herself to the Exempt Fire Company (founded in 1862) for display in their meeting room at Brenham Place.

Lillie died in San Francisco in 1929 aged 86. She bequeathed a third of her fortune for a memorial to the Volunteer Fire Department (1849–66). (The bronze sculpture that was created in 1933 is discussed later.) Lillie also left funds for the beautification of San Francisco. The result was a cylindrical, fluted, 180–feet-high tower with an observation deck at the top. It was designed by architect Henry Howard and built from grey concrete in 1933. The Coit Memorial Tower on Telegraph Hill is now a San Francisco landmark and tourist attraction. Some locals believe the phallic shape of the tower echoes that of the nozzle of a fire hose in order to signify Lillie's association with her beloved fire companies. However, the architect repudiated this interpretation. Inside the Tower there are numerous murals in fresco of the 1930s' social realist variety: they depict the workers who built California's agriculture and industries, banking and law, plus aspects of city and rural life. Readers interested in the art and artists can consult Masha Jewett's history of the Coit Tower while those who wish to learn more about the life of the firebelle herself can read Helen Holdredge's 1967 biography. In Holdredge's view, Lillie became a popular figure in San Francisco because she symbolised the city's 'great love for its volunteer fire companies'. Holdredge also cites a newspaper article that argued Lillie 'personified the woman of the future'. She was indeed a prototype of modern liberated womanhood. Nevertheless, despite the example of Lillie, it was a long time before the San Francisco Fire Department admitted women: the first were recruited in 1987.

## 49 1863

### Unknown photographer, Hand-in-Hand Fire Company No. 1, Philadelphia, with their Steamer

At least one group portrait painting of fire chiefs has been discussed already but the proliferation of group portraits from the middle of the nineteenth century onwards was primarily due to the spread of photography. An early British example, a rather scratched daguerreotype owned by the Library of Congress, was taken in 1850 of members of the Arundel Engine Company standing around a small manual pump. Other daguerreotypes taken in studios and depicting three or more firemen facing the camera, most closely resemble Dutch group portraits

Photographs of American volunteer companies in the street with steam engines, such as the Hand-in-Hand Company No. 1 of Philadelphia, taken in the 1860s often depict them posing informally. In this instance, they smoke cigars and lean on one another and some are out of uniform. Once paid professionals were in the majority, discipline became stricter and the men's uniforms and poses were standardised; consequently, later group portraits tend to be more formal and predictable in their compositions.

50a Leonard Wells Volk, *Volunteer Firefighters' Monument*, (1864). Memorial/public sculpture, marble, granite and limestone, c. 12 metres high. Chicago: Rosehill Cemetery. Photos courtesy of Mary Vileta, oldbones.net

Thousands of such group portraits currently exist. Every local history collection and fire museum possesses examples and they have been reproduced on cigarette cards and picture postcards. Most images show firemen wearing their uniforms and helmets or caps, sitting on, or standing proudly by, their engines and escapes parked in front of firehouses.

In 1914, an unusual photograph was taken of the Fire Department of Gary, Indiana outside the town's central police station, which was situated on a corner. Because the photograph was panoramic, it permits views down two streets and its long width enables one horse-drawn ladder truck and five motorised vehicles and their squads to be represented. Even larger groups of firefighters appear in photographs of bands, parades, inspections, tournaments and funeral marches. Also related to the group portraits – mostly static compositions – are those showing drills and demonstrations of equipment, which naturally show firefighters in action.

Group portraits document fire companies for the benefit of their members, their stations and for posterity. They also emphasize the value of organisation and teamwork, plus the pride officers take in their machines and the professional skills they need to operate increasingly sophisticated equipment.

# 50a/b

**1864**

## Leonard Wells Volk, Volunteer Firefighters' Monument

Volk (1828–95), the sculptor responsible for the firemen's memorial in Chicago's Rosehill Cemetery, founded in 1859, was born in Wellstown, rural New York. He was the son of a marble-cutter active in Pittsfield, Massachusetts during the 1840s. After working for his father, he learnt drawing and clay modelling, and undertook funerary sculpture in St Louis. By 1853, he was in charge of the Rock Island Marble Works, Illinois. On three occasions, he studied sculpture in Rome: 1855–57, 1868–69 and 1871–72. In 1857, he opened a studio in Chicago and nine years later, he and some other Chicago artists founded the Chicago Academy of Design, an art school whose first premises were destroyed in the Great Chicago fire of 1871. (The Academy was the precursor of the now prestigious museum and art school, the Art Institute of Chicago.) Volk is chiefly remembered for making plaster life masks of Abraham Lincoln's face and hands, and for creating a statue of the President, and for designing a tomb for his brother-in-law Senator Stephen A. Douglas. The latter and Volk himself are buried in Rosehill Cemetery.

50b detail.

Chicago's Fireman's Benevolent Association (organised in 1847) commissioned the Monument in memory of all the courageous members of the Volunteer Fire Department. (A paid firefighters' department replaced the latter in 1858.) At ground level, a granite slab records the names of 15 firemen who are buried at the plot. Stone fire hydrants with helmets on top demarcate the Monument's four corners; then there is a rectangular limestone pedestal with panels depicting manual fire engines and a firefighting scene. Above is a drum around which a hosepipe is wrapped. Above that is a tapering, fluted, Doric column made from white marble. On top stands a handsome, young, fireman figure carved from marble. He wears a helmet and uniform and has a speaking trumpet in his left hand; an overcoat is draped over his right shoulder. In 1979, the Monument was refurbished and re-dedicated to the Chicago Firefighters' Union. Memorial services are held at the Monument annually. A replica of it was featured in a funeral scene of *Backdraft*, the 1991 firefighter movie set in Chicago.

## 51 — 1865

## Unknown sculptor, Hand-carved Wooden Statue

51 Unknown sculptor, *Hand-carved Wooden Statue*, (1865). Sculpture, carved and painted wood, 178 cms high. Hudson, New York: American Museum of Firefighting. Photo courtesy of Ed Harvey/Fireservicephotography.com

This life-sized figure is very similar to the earlier wooden statues of Chief Director of the Mechanic Fire Company No. 1, Louisville and of Chief Engineer Harry Howard in terms of the man's clothes, accoutrements, pose and arm gesture. This particular figure was carved from separate pieces of wood, which were then glued together and painted to protect it from weathering. When it was restored in 1957, 60 coats of paint had to be removed! A cavity was also discovered in the chest, which remarkably contained a time capsule with items dating from 1865. During that year the volunteer firefighting companies were disbanded and so the statue was probably made to commemorate the achievements of the volunteers.

Research by curator Karen Del Principe has shown that the statue was associated with the Rutgers' Hose Company No. 26, located at 166 Monroe Street, New York City. (Apparently, the six white stars on the fire chief's red shirt signified a hose company.) The Company was founded in 1839 and its motto was: 'The noblest motive is the public good.' On the shield of the statue's hat, the words 'Chief Engineer' and the initials 'J.G.' appear. If the latter are correct, then the chief engineer referenced was probably James Gulick, who was an intrepid and popular leader of the Volunteer Fire Department from 1831 to 1836. During the 1840s, he was vice-president of the Association of Exempt Firemen. He died in 1861.

Although the sculptor's name is unknown, Del Principe thinks it could have been Charles S. Fordham (1833–1911), a member and secretary to the Hose Company who was also a jeweller and engraver. Alternatively, a ship's carpenter could have been commissioned to create it.

For decades, the statue was mounted on the outside of an old frame house (at first floor level, above the front entrance), Coney Island, Long Island, used by the Gravesend Exempt Fireman's Association, and so it deteriorated badly. Today, the heavily restored and brightly painted statue is one of the prize possessions of the American Museum of Firefighting and its image serves as the institution's logo.

## 52 — 1868

## Helen Paterson, Frontispiece and title page engraved illustrations in Fighting the Flames: A Tale of the London Fire Brigade

Robert Michael Ballantyne (1825–94), the Scottish-born author of the novel *Fighting the Flames*, was a prolific Victorian writer of popular adventure books for boys. He himself had experienced adventure in foreign lands

52 Helen Paterson, Frontispiece and title page engraved illustrations in *Fighting the Flames: A Tale of the London Fire Brigade*, (London: James Nisbet & Co., 1868). Author's collection

because in his youth he had worked for the Hudson Bay Company in the wilds of Canada. Sir Walter Scott's romantic novels influenced his writings and Robert Louis Stevenson in turn admired him. *Fighting the Flames*, which seems to have been the first literary mythologising of the work of the Metropolitan Fire Brigade, was a commercial success because it sold thousands of copies and was subsequently issued in new editions by several publishers well into the twentieth century. Various artists illustrated the different editions, which at first were monochrome but later became full colour.

Ballantyne's descriptions of fires and firefighting methods have a documentary quality and indeed one of his characters is the real fire chief James Braidwood. The book's climatic chapter concerns the infamous 1861 Tooley Street fire at which Braidwood was killed. In his preface, Ballantyne thanks Captain Shaw (Braidwood's successor), Sampson Low (Secretary of the London Fire-Escape Society) and other fire officers for helping him to attend fires, and so it is clear Ballantyne researched his subject thoroughly before setting pen to paper. He also describes two weeks spent with the MFB in his later book *Life in the Red Brigade* (1873 and 1887).

Helen Paterson (1848–1926, later Helen Allingham), a book and magazine illustrator, and watercolour painter of idealised landscapes, was born into a middle-class family in the Midlands. Along with her friend and fellow illustrator Kate Greenaway, she became one of the most successful female graphic artists of the late nineteenth century. She was one of first women to become a member of the Royal Watercolour Society and is best known for illustrating the serialised novels of Thomas Hardy. Paterson trained at the Birmingham School of Design and then, in 1867, she moved to London to continue her studies at the Female School of Art and the Royal Academy Schools. At the latter, she was helped and encouraged by Millais and so it is likely she knew his 1855 fireman painting. The engraved illustrations for the publisher James Nisbet, therefore, were among her earliest commissions.

The motifs of the two images reproduced – a resourceful fireman rescuing a mother and infant (with his teeth!) via a ladder, and a fire engine dashing through the streets – are, of course, highly familiar. There are four more illustrations in the book all set in interiors such as fire stations and bedrooms.

## 53 1870

## Unknown artist, Sapeurs-Pompiers (Instruction Militaire)

Épinal, a town in the Vosges region of France, is famous for its paper industry and its popular prints, especially those produced by Jean-Charles Pellerin (1756–1836) during the nineteenth century. His company Fabrique de Pellerin was founded in 1796. Bright colours were added to wood engraved or lithographed images and peddlers used to sell the visually attractive prints to adults and children around the country. This particular print is instructional or educational in that it documents the uniforms, ranks, equipment and rescue deeds of Parisian fire brigades. Four bands of images plus brief captions depict a variety of scenes in a manner resembling a modern comic strip. Since a steam engine appears in one vignette, the print probably dates from

53 Unknown artist, *Sapeurs-Pompiers (Instruction Militaire)*, (1870?). Print, No. 222. Épinal: Pellerin & C

54 Thomas Waterman Wood, *The Rescue*, (1870). Oil painting. New York: Collection of The New York City Fire Museum

around 1870. (Steam engines were introduced during the 1860s.)

Urban artists and intellectuals fascinated by French folk culture valued popular prints. Courbet was one leading French painter whose compositions were sometimes inspired by their imagery, but in this case the influence was surely the other way round: the departure scene in the centre of the top band closely resembles that of Courbet's 1850–51 canvas *The Departure of the Fire Brigade*.

## 54 1870

## Thomas Waterman Wood, The Rescue

This painting depicts a night scene in which a bearded, red-shirted American fireman, who is probably a member of a 'Hook and Ladder' company, carefully descends a ladder while holding a young girl who gazes fearfully back towards the fire that threatened her. (The steps of ladders made slippery by water were especially hazardous.) A speaking trumpet dangling from the fireman's right arm seems likely to impede his progress. He

55 Unknown engraver, *The Brave Fireman*, (1871). Engraving, 27.5 x 22.5 cms, published on the cover of *The Graphic*, 28 October 1871. Author's collection

wears a leather hat with an extended rear brim – probably the model designed by Henry Gratacap in 1828 – that is also ribbed for extra strength. Light from the fire illuminates the two figures set against a dark background. Burning embers cascade down and the high distance of the couple from the ground is conveyed by the low position of a building resembling a city hall in the background.

Wood (1823–1903), the son of a Montpelier (Vermont) cabinet-maker, was an itinerant portrait, landscape and genre painter, and expert etcher who taught himself largely by copying the old masters and by working from nature. He travelled extensively to Boston, MA, Canada, London, Paris and Italy and then established a studio in New York. Eventually, he became president of the National Academy of Design. Wood had a realist agenda and via his anecdotal genre scenes – some depicting African-Americans – studies of American characters, and prints sought a popular audience. Indeed, some scholars regard him as a precursor of Norman Rockwell. More than 200 of his oils, watercolours and prints are preserved in the Wood Art Gallery, Montpelier.

*The Rescue* seems to be the only work by Wood depicting firefighting, so one presumes the subject was of momentary interest. This image became more widely known via reproduction when, on 8 February 1873, *Harper's Weekly* magazine published on its cover a black-and-white wood engraving of it by the artist. Today, so-called 'Art Prints' of *The Rescue*, that is, prints hand-coloured with watercolour pigments to give them the appearance of paintings, are being marketed via virtual galleries such as AllPosters.com and ArtCity-com.

## 55 — 1871

## Unknown engraver, The Brave Fireman

In the second half of the nineteenth century, paintings or photographs of firemen were often translated into wood engravings so that prints could appear in British and American popular illustrated magazines such as *The Graphic*, *The Illustrated London News* and *Harper's Weekly*. Alternatively, similar images were created specially for such magazines.

During the night of 7 October 1871, a fire occurred above a chemist's shop in Gray's Inn Road, London. With the help of a police officer, a fireman called Joseph Ford wheeled his escape ladder to the premises and ran it up to a top window. Repeatedly, he climbed the ladder to rescue five people. He then heard a woman's scream. Despite his exhaustion, Ford climbed again and managed to grasp the woman. On the way down, she fell but was caught by a crowd waiting below. Ford's belt then became entangled in the mesh of the chute attached to the underside of the ladder and he was severely burned. He crashed to the ground and died shortly afterwards in hospital.

Three weeks later, *The Graphic* (1869–1932) paid homage to Ford with a cover image showing him descending a ladder carrying an unconscious female. The ladder he is using is being stabilised by two ropes and the net chute underneath is clearly visible. The creator of the engraving is not credited but may well have been Charles Green (see illus 60). What is unusual about this engraving is the almost abstract, rococo-like design of the flames and clouds of smoke rising from beneath and swirling around the couple. In a short text about the incident, Ford is compared to heroes of ancient times – 'mighty men before Agamemnon' – and his deed and death are celebrated and lamented in a poem entitled 'The Brave Fireman' written by Arthur Locker, editor of *The Graphic*.

At this period, British firemen were poorly paid and worked very long hours. Pensions in the event of injury or death were also difficult to obtain. Ford had been married with children. His employers – the Metropolitan Board of Works – grudgingly awarded his widow a pension of one pound per week. However, the press and the public were so impressed by Ford's bravery and sacrifice that they raised £1000 for his widow; whereupon, the miserly Board of Works cancelled her pension!

56 Victor Pierson & Paul Poincy, *Volunteer Firemen's Parade, March 4th 1872*, (1872). Painting, oil on canvas, 183 x 274 cms. New Orleans: Courtesy of the Louisiana Historical Society, Collection of the Louisiana State Museum

## 56 1872

### Victor Pierson & Paul Poincy, Volunteer Firemen's Parade, March 4th 1872

This remarkably ambitious group portrait and townscape depicts the 1872 annual parade of the volunteer fire companies of New Orleans. They gather around a statue of statesman Henry Clay, which was then located at Canal and Royal Streets. Most firemen wear red shirts and fire hats, and are on foot but others and local dignitaries are on horseback. On the right is a steam fire engine drawn by four horses while in the foreground an African-American woman sells refreshments to a queue of firemen. Of course, the fire companies consisted of white men because of racial segregation. Crowds of spectators gaze upon the scene of civic pride from the balconies of nearby buildings.

Firemen's parades still occur but during the nineteenth century, they were frequent and popular: fire companies across America delighted in the pageantry of parades, which often involved welcoming companies that were visiting from other towns. Banners, bunting, flags, floral decorations, music from bands and singing accompanied processions, and at night there were fireworks, lanterns and torches.

This painting may have been influenced by the Charleston work by Mayr discussed earlier. It required the collaboration of two painters – Paul Poincy and Victor Pierson – and one photographer – Theodore Lillenthal. Poincy (1833–1909) was a New Orleans artist and art teacher of French descent who had studied in Paris at the École des Beaux Arts and the private Académies of Gleyre and Julian.

He became noted for his landscapes, street scenes, portraits and religious paintings. Pierson, a British artist skilled at portraiture, hunting scenes and depicting animals such as horses, arrived in New Orleans in 1868 and remained until the mid–1870s. Lillenthal (1829–86), a local studio photographer of German origin, provided the two painters with numerous portrait photographs, which helped them to portray over 120 identifiable individuals who included a Grand Marshall, an orator, the Chief of Police, a judge and spectators besides fire chiefs and firemen.

Large group portraits of this kind are notoriously difficult because of the need to do justice to the faces of all the individuals while at the same time integrating them into the composition as a whole. Integration is not fully achieved but the characterisation of individual countenances is often excellent. The mural-size canvas was first exhibited in February 1873 and hung for many years in the fire commissioner's office, City Hall, New Orleans.

## 57 1874

## George Halse, The Fireman

This group sculpture depicting a bearded British fireman rescuing a naked baby and a swooning girl was exhibited at the Royal Academy in 1874, consequently it must date from that year or shortly before. George Frederick Halse (1826–95) was a minor Victorian sculptor, illustrator, poet, novelist and author of a book about clay modelling and the principles of sculpture. It seems he was a self-taught, spare-time artist (he worked for 39 years in a bank). As a sculptor, he specialised in bronzes and stone carvings in the round and in relief, depicting such subjects as children, family groups, portrait busts, Roman soldiers in battle and allegorical figures, all rendered in a neo-classical style. In this instance, it would seem that, in response to the new realist art of the period, he was prompted to tackle a contemporary subject. Even so, the work embodies the values of 'Ideal sculpture' in which raw nature 'is perfected' by the artist. Often Halse would model a sculpture in clay, then make a plaster cast, which would be copied in marble by a mason; Halse would then add the finishing touches. Although clearly a skilful academic sculptor, most of Halse's work is rather insipid and sentimental in character.

Halse's fireman steps forward with his left leg and turns his head to his right as if conscious of the fire from which he has emerged with the two victims. (Of course, a disadvantage of freestanding sculptures is that an environmental context cannot be provided only suggested.) An axe lies discarded at his feet. The fireman's alert, determined face has a particularly fine profile and with his crested helmet he resembles the Roman soldiers Halse enjoyed depicting. A photograph of the sculpture is preserved in an album owned by his descendants. Below the photograph is a short poem, written by Halse himself, which reads:

> Heroes there are, and unrecorded deeds
> Of valour in the field of humble life;
> And noble, nameless men whom no one heeds,
> Fighting in other than battle's strife;
> Most swift when e'er the wave of anguish pleads;
> Most resolute where danger is most rife;
> In duty stern, yet, rugged though they be,
> All-gentle:- Nature's very chivalry.

It is clear from the sculpture and poem that Halse aimed to represent a modern, everyday hero exemplifying the ideals of manliness, nobility and bravery. Victorian works similar to this, when produced in editions, could be displayed as ornaments in many middle class homes. They were the forerunners of the dozens of firefighter figurines currently on the market.

57 George Halse, *The Fireman*, (1874?). Sculpture, marble, 96 cms high. Collection of Benedict Read. Photo courtesy of the Conway Library, Courtauld Institute of Art, London and Elizabeth Farningham

## 58 1877

## Melzar Hunt Mosman, Fireman's Monument

In 1869, the Firemen's Benevolent Association of New Haven, Connecticut began to raise $5000 to fund a monument in memory of New Haven's deceased firemen or 'soldiers of peace' to be located in the Firemen's Lot, Evergreen Cemetery. A fire department had existed in New Haven since 1784. The sculptor they commissioned to create a life-size bronze figure to crown a tall stone pedestal was Melzar Hunt Mosman (or Mossmann, 1846–1926) of the Ames Sword & Bronze Manufacturing Company, Chicopee, Massachusetts, where his father Silas was employed as a foundry worker. In 1867, Mosman travelled to Paris in order to develop his modelling and casting skills. During his artistic career, he undertook commissions for Civil War monuments and equestrian statues.

Mosman's fireman is an elegantly posed, alert figure who sports a moustache and goatee beard and wears a tall fire hat, shirt, trousers and boots. He steps forward with his left leg and gazes toward New Haven's City

58 Melzar Hunt Mosman, *Fireman's Monument*, (1877). Memorial/ public sculpture, bronze, stone and granite, approx 7 m high. New Haven, Connecticut: Evergreen Cemetery. Photo © Edward F. Flynn Jr

Hall. In his right hand, he holds a wrench and over his left forearm is draped a coat. The base and pedestal, decorated with bas-relief bronze plaques featuring New Haven's coat-of-arms, ivy leaves, fire engines, firefighting apparatus and emblems, is also particularly fine.

On 9 July 1877, a day of heavy rain, the Monument was unveiled and dedicated in a ceremony that involved music, prayers and speeches, and it was re-dedicated on 16 October 1993 following a restoration. The bronze had turned green with age and so Tom Laugeni, an ex-volunteer firefighter and head of A. Laugeni & Sons, Inc., of West Haven, refurbished it free of charge.

There is another fireman's monument in Evergreen Cemetery near to Mosman's. It is a smaller stone structure with a bas-relief carved figure of a fireman holding a hosepipe set inside a niche, which is surmounted by the mythical bird Phoenix rising from ashes. Phoenix Engine Company No. 5 erected it in memory of Bevil Sperry, one of its members, who was killed on duty in 1854. Thomas Phillips (1805–89) of Thomas Phillips & Son (a New Haven marble works) a stonemason born in Plymouth, England carved the relief, which is now much weathered.

## 59 — 1880

## H.S. Hebard, Fireman's Monument

While many of the fireman statues in cemeteries are generic – representing the whole body of firefighters (as in the case of Launitz's and Volk's sculptures) – others represent, or are named after, particular individuals. For example, there is a tall, Egyptian-style monument in the

59 H.S. Hebard, *Fireman's Monument*, (1880). Memorial, St Johnsbury granite, 15 m high. Rochester, NY: Mt Hope Cemetery. Photo: Richard Margolis, Rochester

60 Charles Green, *Heads of the People: The Fireman*, (1880). Wood engraving, 23 x 30 cms. Published in a supplement to *The Graphic*, 10 January 1880. Author's collection

Firemen's Plot of Mt Hope Cemetery, Rochester, New York, which commemorates Wendell Charles Byers, the first chief of the Rochester Fire Department. It was commissioned by the Rochester Firefighters Benevolent Association, cost $8000, and was dedicated on 9 September 1880. The sculptor H.S. Hebard, an ex-volunteer fireman, executed the design and his Steam Marble Works fabricated it. Byers, a dignified moustachioed individual, is dressed in uniform, belt and helmet, and carries a coat over his left arm. His relaxed pose implies that he is resting while returning from duty. A relief showing a winged orb, symbolising Time and Eternity, decorates the Monument's square shaft.

Some further examples: the Bevil Sperry Monument in New Haven has already been described and in Collyer Park, Pawtucket, Rhode Island there is a bronze portrait of Samuel S. Collyer, Chief Engineer of the Pawtucket Fire Department who died in 1884 from injuries received when a fire engine overturned while answering an alarm. The statue, mounted on a granite base that has a bronze relief depicting the accident, was modelled by the artist Charles Dowler and completed in 1890. Another memorial depicts Fire Chief Philip Eckel, who died on duty in 1886. It is made from granite, dates from 1900, and is located in Fayette Park, Syracuse, New York.

## 60 — 1880

## Charles Green, Heads of the People: The Fireman

In January 1880, *The Graphic* published an engraving by Charles Green, a prolific Victorian illustrator of magazines and books by Dickens and others, depicting a bearded British fireman wearing a brass helmet and holding a hose nozzle. Firefighters assigned to fire engines with pumps and hoses take pride in being first to the seat of fires and in taking charge of hose nozzles. (In the medium of photography, the hose shot was to become perhaps the most common motif of firefighting iconography.) Light from an unseen fire illuminates his face and burning embers float around his head. What Green's graphic anticipated, in effect, was the movie close-up.

A reviewer in the journal *The Fireman* (1877–1906), quoted by historian Sally Holloway, was impressed and claimed that the man's features were 'animated with an expression of unmistakable intelligence and heroic resolve which suggests he is prepared for any conceivable act of intrepid daring or self-sacrifice. A spirit of sympathetic humanity permeates every line of that face.' Such images, the writer thought, could inspire beholders to accomplish noble deeds.

Green's engraving was one of a series entitled 'Heads of the People', which the Dutch painter Vincent van Gogh admired. Vincent lived and worked in London from 1873 to 1875 and was an avid reader and collector of British illustrated magazines; he was also moved by sentimental Victorian narrative paintings. 'Heads of the People' later inspired him to produce similar portraits of doctors, fishermen, miners, peasants, postmen, soldiers and weavers. If Vincent had known a fireman, no doubt he would have painted his portrait too.

## 61 — 1880

## Samuel Calvert, Bush Fire in Australia

## 62 — 1885

## Unknown artist, Fighting the Fire

Samuel Calvert (1828–1913) was an English oil and watercolour painter, lithographer and wood engraver who lived and worked for many decades in Australia. Born in Brixton, London, he sailed to Adelaide in 1848 and then, in 1852, he moved to Melbourne. Calvert came from an artistic family: his father Edward was a painter and engraver, a follower of William Blake and a friend of Samuel Palmer. Calvert, who learned wood engraving from his father, proved to be a prolific and successful artist who generated hundreds of illustrations of Australian subjects for advertisements, books, magazines and newspapers, and designs for postage stamps.

His 1880 print *Bush Fire in Australia* depicts a dramatic, indeed melodramatic scene, in which no less than three fires threaten farms in the countryside. The largest fire, on the horizon, is already out of control and is about to engulf some farm buildings. A stockman on horseback and a herd of cattle flee in terror over a fence. Simultaneously, in the foreground, three male farmers or labourers attempt to suppress a grass fire, which they may have started to create a fire-break, close to their homestead. The latter is made of wood and so will burn readily. Two men are draping sheets – probably soaked in water – over it to help protect it from sparks. On the right are two women and a child in an agitated state. Although predominantly brown in colour, this print vividly conveys the dangers posed by bushfires.

The second illustration (black-and-white, by an uncredited artist), which was published in a monthly newspaper issued by Melbourne publisher Alfred Martin Ebsworth, depicts another rural scene in which tough-looking bushmen attempt to beat out a fire by using branches. In the foreground, a young woman dressed in a peasant-like costume and represented in profile offers a cup of water to a bearded perspiring labourer who is represented from the front; the calmness and stasis of

62 Unknown artist, *Fighting the Fire*, (1885). Wood engraving illustration published in a supplement to *Australasian Sketcher*, 8 April 1885. Photo: Melbourne, Australia: La Trobe Picture Collection, State Library of Victoria

61 Samuel Calvert, *Bush Fire in Australia*, (1880). Print/colour wood engraving, 32.9 x 45.8 cms. Canberra: Print Collection, National Gallery of Australia. © NGA

63 Unknown artist, *French Fireboy*, (n.d.). Trade card, 7 x 9 cms. Boston, MA: Jordan Marsh & Co. Author's collection

the two central figures contrasts sharply with the animated figures, smoke, flames and danger behind them. No story accompanied the image and so it was not linked to any particular contemporary event. However, devastating bushfires had occurred in the state of Victoria four years earlier on so-called 'Black Thursday', that is, 6 February 1881.

Bushfires, started either by natural or human means, are commonplace in Australia even today when many fire wardens, rural fire brigades (with 70,000 volunteers) and aircraft with water tanks exist to combat them. Severe bushfires have killed many Australians, burned farms, villages and suburbs, caused air pollution and have even threatened to destroy parts of major cities. Contemporary firefighters report that it is not only the tremendous heat and smoke of bushfires that make them nerve wracking but also the extremely loud noise they generate. Still, silent pictures, of course, cannot convey this characteristic of such fires.

Bushfires are not confined to Australia. They regularly occur in California and even within sight of New York: on 20 April 1963 a massive bushfire on Staten Island occupied 1500 firefighters for ten hours and destroyed over 100 homes.

**63**

**Unknown artist, French Fireboy**

**64** **1880s**

**Unknown artist, Paris Fireman French Army 1886**

**65** **1880s**

**Unknown artist, Running Fireman**

**66** **1881**

**Unknown artist, The Fireman**

**67** **1880s**

**Unknown artist, The Fireman**

Trade cards, which were a form of advertising and printed ephemera, had their origin in the late eighteenth century but did not become widespread until the advent of colour lithography during the 1870s. At the height of their popularity, they were small coloured images with slogans or captions printed on cardboard mounts that promoted local merchants, manufactured goods and services of all kinds. Some accompanied wrapped products while others were available in shops or were handed out in the street. Some images were unique to a particular company and so depicted its products but other cards were generic in character and so their designers could treat almost any subject.

In terms of themes depicted, trades, professions or

64 (left) Unknown artist, *Paris Fireman French Army 1886*, (c. 1880s). Trade card, 3.5 x 7 cms. New York: Kinney Tobacco Company. Author's collection

65 (centre) Unknown artist, *Running Fireman*, (c. 1880s). Trade Card, 13.8 x 7.7 cms. Publisher: Gies & Co., Buffalo, New York.

66 (right) Unknown artist, *The Fireman*, (1881). Trade Card, 11.3 x 7.4 cms. Publisher: Cosack & Clark. Photo: Trade Card Place

occupations proved attractive and so became commonplace. This interest seems to have been a response to the increasing complexity of industrial society with its high division of labour that generated a multiplicity of occupations that demanded to be documented and classified. Children being educated are encouraged to learn about occupations to gain an overview of society and because they will soon have to choose one.

Pictures of firemen were thus disseminated by American companies such as Babcock, which manufactured chemical fire extinguishers. In England, a chromolithographic scrap depicting a metropolitan fireman up a ladder with his axe raised is preserved in the 'Trades and Professions' section of the John Johnson Collection of Printed Ephemera, Bodleian Library, University of Oxford. It was printed in order to be pasted down in the scrapbooks or albums, which were so popular with the British and American publics – especially women and children. Trade cards and scrap images encouraged the urge to collect. Examples of American scrapbooks are preserved in Duke University's Rare Book, Manuscript and Special Collections Library.

One undated trade card depicts a cherubic fireboy wearing the uniform of a *sapeur-pompier* and holding a hosepipe. The figure is placed against a shiny gold ground. At the top of this small charming card is printed 'Jordan Marsh & Co.' the name of Boston's most famous department store. Another, even smaller image of a French *sapeur-pompier* wearing a uniform of 1886, appeared in a set of cards devoted to military and naval uniforms issued by the Kinney Tobacco Company of New York. The card was included in packs of Sweet Caporal cigarettes and dates, presumably, from the late 1880s.

Cigarette cards are perhaps the best-known type of trade card. The oblong cardboard inserts were originally intended to strengthen cigarette packets. Another Kinney cigarette card from the 1880s, entitled *A Deadly Peril*, had a monochrome image of a young woman standing by an open window being threatened by smoke and flames. This was a semi-transparent 'magic changing card' because when held up to the light a second coloured image appeared featuring a fireman on a ladder rescuing the female. A very similar image appeared on a card, entitled *The Fireman*, distributed in packs of Duke's Cigarettes (W. Duke, Sons & Co., Durham, North Carolina) during the period 1885–95. The card was one of a series of 50 depicting 'Perilous Occupations'.

The names of the artists who created the images on trade cards do not seem to be known but some of the commissioning companies are; for instance, Gies & Co., a printing firm located in Buffalo, New York, issued a set of 24 cards depicting different occupations. All the images were of men engaged in their activity plus a related object in an upper corner. In this instance, a fireman dressed in red shirt and blue trousers and carrying a trumpet is shown running towards a fire. The object in the top left-hand corner is a lamp. Acorn Stoves and Ranges was one company which used the card to promote its products.

67 Unknown artist, *The Fireman*, (1880s?). Chromolithograph, die cut, embossed scrap, 19 x 10 cms. Author's collection

Another American firm – Cosack & Clark – published, in 1881, a set of humorous, cartoon-like images, entitled *The Fireman*, depicting six scenes from his daily life. The one illustrated showing a moon-faced, red-shirted fireman carrying no less than five members of a family down a ladder was clearly parodying the popular rescue theme. In their rendering of figures, trade card artists frequently employed caricature and satire.

As several books and many websites on the Internet indicate, antique trade cards are currently collected by enthusiasts interested in history, advertising, consumerism and popular culture, and rare cards fetch high prices at auction. Furthermore, trade cards are still being designed and issued. Kingsley Barham of Chestnut Publications, Delray Beach, Florida, for example, has published cards about marijuana, motorcycles and tattoos. In 2002, he decided to produce a set entitled 'The Heroes of the World Trade Center', that is, 202 palm-size cards with photo-portraits and personal details of 170 firefighters and other victims of 9/11. Barham claimed that his motives were patriotic and charitable, and that he wanted to give the victims a human face. The negative public reaction to this commercialisation of the event was so great that Barham temporarily ceased publication. However, 15 prototypes had been printed, including one about New York fireman Manuelle DelValle Jr., and circulated to retailers and card collectors. Examples of these have been offered for sale via Internet auction websites. As one journalist pointed out, in publishing the memorial cards Barham was exercising his right of free speech under the First Amendment and that his critics seemed to be upset by the format – trading cards – rather than by their content.

68 Thomas B. Worth, *Darktown Fire Brigade. The Chief, on Duty*, (1885). Hand-coloured lithograph, 41 x 31 cms, published by Currier & Ives, New York. Photo: Washington, DC: Library of Congress, Prints & Photographs Division

## 68 — 1885

**Thomas B. Worth, Darktown Fire Brigade. The Chief, on Duty**

## 69 — 1887

**Thomas B. Worth, Darktown Fire Brigade. Hook and Ladder Gymnastics**

A second series of Currier & Ives prints treating the subject of firefighting featured the so-called 'Darktown Fire Brigade', the images for which were created by the artist Thomas B. Worth (1834–1917). This Brigade, of course, was fictional and 'Darktown' was a humorous version of the ghettos in which most African-Americans lived during the nineteenth century. About 200 prints were issued about life in Darktown between the 1870s and the 1890s – the Library of Congress has 197 prints and the Museum of the City of New York has 172 – but only a dozen of these were about firefighting.

A typical example, dated 1885, is a caricature of a pompous, overweight, black Fire Chief dressed in uniform and helmet, holding a speaking trumpet and shouting 'Lite up dem hose dar-yous heah me!' The Chief's face reiterates white racist perceptions of African-Americans: thick lips, large nose and bulging eyes. Behind him, two teams carry ladders and operate a hand pump. A fire is taking place in a wooden house and figures in absurd postures are escaping via the windows.

Another example, dated 1887, again depicts the Darktown Brigade – all attired in red shirts and blue trousers – tackling a house fire. On the left, two firemen struggle to prevent an unbalanced ladder, which has a victim in a white nightshirt clinging to its top, from toppling over. The victim is in danger of having her bottom pierced by the point of a hook held by a fireman standing on a ladder carriage. A caption reads: 'Brace her up dar! and cotch her on de fly!' In the right-hand corner, two firemen direct their hoses towards the house, but the water pressure is so feeble the jets droop pathetically far short of their target – the hoses are leaking water due to holes. In the background, two more firemen are waving and blowing their trumpets furiously. Amid this scene of mayhem, a terrified cat flies through the air.

The general impression conveyed by such prints is that fires in the segregated communities were comical affairs and that their firemen were bumbling fools. Marcy S. Sacks, a scholar who has analysed the prints in order to confront racism, argues that they 'used caricature and coarse stereotypes to lampoon blacks who moved north after the Civil War'. Robyn Cooper has also claimed that the images were 'an affirmation of the racial superiority of the white Caucasian fireman'.

Of course, the prints were made for, and were

69 Thomas B. Worth, *Darktown Fire Brigade. Hook and Ladder Gymnastics*, (1887). Hand-coloured lithograph published by Currier & Ives, New York. Photo: Washington, DC: Library of Congress, Prints & Photographs Division

bought by, middle-class and lower-middle class whites to decorate their homes. One memoir of the period describes a home in which a whole room had been pasted with Darktown prints from floor to ceiling. An American company called C.W. Briggs Co., active from 1868 to 1930, also issued lantern-slides of Darktown Fire Brigade prints. In addition, prints decorated the inside lids of cigar boxes and formed the basis for children's picture puzzles manufactured by Parker Brothers.

Defenders of Nathaniel Currier and James Merritt Ives claim that the Darktown prints were light-hearted rather than malicious, and point out that both men supported the abolition of slavery. Worth himself was a genre painter of people, horses and landscapes as well as being Currier & Ives' principal caricaturist. Perhaps he was not a racist because one of his genre paintings, dated 1868, is a sympathetic portrayal of a black fiddle player adjusting his cravat.

One contemporary value of the Darktown prints is that they inform viewers that black firefighters did exist in nineteenth-century America. Furthermore, in the case of the 1885 print, this can be regarded as an early indication of the way in which fire chiefs, even white ones, were soon to become comic figures. Interestingly, there were all-black fire companies formed in the United States during the early nineteenth century but they tended to be neglected or actively suppressed by rival white companies. During the 1820s, the Savannah Fire Company, which was managed by white officers, recruited 'free men of color, free Negroes and hired slaves'. Some white volunteer companies also tolerated African-Americans as drivers because of their skill with horses or permitted them to undertake menial tasks such as dusting. Then, during the period 1850 to 1960, black fire companies existed as part of segregated fire departments. There are several contenders for the first African-American to achieve the position of fire chief. One candidate is Patrick H. Raymond (1831–94), who was appointed Chief Engineer of the Cambridge Fire Department in 1871. In Los Angeles today, an African-

American Firefighter Museum documents the history of their achievements and struggle. It opened in 1997 and occupies a building that used to be the firehouse of one of only two all-black companies that operated in LA for 75 years.

## 70 — 1885

### R.A. Fox and J.A. Fraser Jr, Going to the Fire

## 71 — 1906

### Charles Edward Stewart, Fire!

By the early nineteenth century, specially trained horses (teams of two, three or four) were being used to pull fire appliances and by the late 1850s, manual pumps began to be replaced by steam-powered ones, which were operated and maintained by firemen designated engineers. (Steamers dated back to 1829 but did not come into general use until much later.) Horse-drawn steam engines, which often look as if they themselves are on fire (belching sparks, steam and black smoke), speeding through the streets provided an exciting spectacle but they were also noisy and dangerous. The dangers were many: spectators could be run over (an image published in *Harper's Weekly* on 24 February 1872 shows spectators scattering in all directions; a similar scene in *Frank Leslie's Illustrated Newspaper* on 10 October 1874 showed a fireman running in front of the horses to warn people to move aside); the horses could fall or crash thus injuring themselves and the men on board; the steam engines could explode: an image in *Frank Leslie's Illustrated Newspaper* on 4 July 1868 depicted such an explosion that took place in the crowded street outside New York's Bowery Theatre.

70 R.A. Fox and J.A. Fraser Jr, *Going to the Fire*, (1885). Painting, oil on canvas, 162.5 x 117 cms. Philadelphia: Historical Collection of the Insurance Company of North America

As explained earlier, when representing speeding fire engines artists had favoured a side view. Later, however, artists such as R. Caton Woodville (1856–1927) realised that a three-quarter view was better, witness his 1882 *Illustrated London News* print *Fire!* Even more dramatic was the frontal view because then the horses appear to be bearing down on the viewer. Hence, the near-frontal view preferred by the artists R.A. Fox and J.A. Fraser Jr in their 1885 oil painting *Going to the Fire*. (Appropriately, a painting of two horses and two firemen was painted by two artists.) Their rendition is somewhat naïve but does convey the strength and speed of the horses; the determination of the bareheaded driver and the moustachioed engineer riding the engine; and the latter's mechanical power. A suction pipe mounted on the left side of the engine looks to modern eyes like a rocket launcher. A lighted lamp hanging from the engine confirms it is a night scene.

Robert Atkinson Fox (1860–1935), a highly prolific artist and popular printmaker, was a Canadian who also worked in the United States. His prints depicted a wide range of subjects – landscapes, gardens, people, animals, etc. – consequently, firefighting was not a specialist interest. Fox experts claim he was a fast and facile painter who often worked from memory and the imagination. Today, however, American collectors seek after his prints and there is a R. Atkinson Fox Society devoted to him.

Less is known about his collaborator John Arthur Fraser (1838–98), who was born in England and worked in Canada. Fraser was a painter of Canadian landscapes, a teacher, illustrator and tinter of portrait photographs. Both artists lived for a time in Toronto, so it is probable their collaboration occurred there.

In England, Charles Edward Stewart, a Glasgow-born artist who trained at Royal Academy School in London and the Académie Julian in Paris, and who exhibited pictures of rustic and historical scenes at the Royal Academy from 1887 to 1938, painted a very similar picture to that of Fox and Fraser, in 1906. It shows three metro-

71 Charles Edward Stewart, *Fire!* (1906). Painting. London: Chartered Insurance Institute

politan fireman aboard a steam fire engine being pulled by a pair of galloping white horses. This painting is executed with considerable verve.

The near-frontal view Fox, Fraser and Stewart adopted later became a pictorial cliché especially in the medium of photography. For instance, on 23 June 1910 Delmar 'Barney' Roos, a photographer employed by the New York agency Brown Brothers, was in New Haven, Connecticut on assignment when he heard a fire bell ringing. Quickly setting up his glass plate camera, he managed to capture a shot of a steamer manned by two firefighters and pulled by three horses coming towards him at full pelt. This often-reproduced image was later dubbed 'the greatest fire engine photograph of all time'.

## 72 — 1886

## Tsukioka Yoshitoshi, Moon and Smoke

This exquisitely decorative Japanese print was one of a set entitled *One Hundred Aspects of the Moon* created by Yoshitoshi (aka Yonejiro and Yoshitoshi Taisio) between 1885 and 1892, which proved highly popular. The artist, who lived from 1839 to 1892, resided in the city of Edo (renamed Tokyo in 1868) and is now regarded as the most important printmaker of the Meiji period (1868–1912). Yoshitoshi, the son of a merchant, was trained at the school of Kuniyoshi and although he later experienced poverty and mental illness, made a career as a painter, printmaker and art teacher. Several newspapers and book publishers commissioned illustrations from him. He lived through bloody, politically turbulent times and witnessed the forced Westernisation of Japan. In his art, he sought to preserve traditional Japanese values while tackling both historic and contemporary subjects.

Edo suffered many serious fires because its buildings were made from wood and paper, and heated by charcoal braziers, and so its rulers ordered citizens to maintain barrels and pails of water, and established fire patrols and numerous fire brigades each of which was identified by a letter from the Japanese alphabet. The print depicts two firemen – hikeshi – standing on the roofs of buildings while flames engulf the space between them. Both are holding 'matoi', that is, standards that identified the different fire companies and marked assembly points. Matoi derived from military standards or banners and consisted of a wooden pole with geo-

72 Tsukioka Yoshitoshi, *Moon and Smoke*, (1886). Woodblock print, 35.7 x 24.7 cms. Publisher Akiyama Buemon, Tokyo. Reproduced courtesy of the Museum of International Folk Art, Museum of New Mexico, Santa Fe

metric forms at the top plus streaming ribbons that revealed the direction of the wind. (Matoi are still carried in modern street festivals.) Fire wardens in tall watchtowers would beat drums or planks, or ring bells to sound the alarm and rival fire companies would dash to fires in order to claim the right and the fee to fight them. The winner would plant its standard on the roof and begin work, which was often a rapid demolition of buildings in order to prevent the flames from spreading. Brigades were equipped with axes, hooks and mallets to tear down buildings and with wooden 'dragon pumps' but their water jets had a limited reach because they lacked hoses.

Like the American fire laddies of the mid-nineteenth century, hikeshi had the reputation of being fearless but also rowdy, violent drunkards. Indeed, some citizens considered them more dangerous than the fires they fought! Others regarded them as popular heroes. Fistfights occurred between rival companies and with other groups such as sumo wrestlers. In 1886, Yoshitoshi created a diptych depicting such a battle that had taken place all day between fireman and wrestlers at the Shinmei Shrine, Edo in 1805. Such incidents inspired plays in kabuki theatres and so there are prints depicting actors dressed as firemen holding their matoi. Yoshitoshi and Kunichika Toyohara (1835–1900) created examples.

Like so many Japanese, the hikeshi – who by the early nineteenth century were mainly drawn from commoners employed in the building trades – liked to have their bodies heavily tattooed to demonstrate their masculinity, their solidarity with one another and their ability to endure pain. They also thought the tattoos would protect them from danger. Popular designs were water dragons, coy fish and water symbols.

The large figure on the left, a matoimochi (standard bearer) is dressed from head-to-toe in protective clothing – hood, coat, trousers and gloves – with bold patterns and Japanese characters on their surfaces. Many firemen's ceremonial wide-sleeved jackets were elaborated decorated with tassels, embroidery and motifs such as water and the Phoenix bird.

The protective clothing, called 'hantan', was made from thick, multi-layered cotton. It shielded the body from falling debris and was wetted before or while the fireman was exposed to heat.

Some hikeshi were called 'tobi' because they worked at heights like western roofers and steeplejacks. They employed lightweight ladders made from bamboo and developed acrobatic skills in using them that are recorded in a number of early photographs and in a print by Kunitoshi Utagawa (1838–98). Such gymnastic displays at the tops of ladders were, and still are, demonstrated in the annual New Year parades Japanese fire brigades are so fond of holding.

73 John Tenniel, *The "Fire Fiend"*, (1887). Cartoon published in *Punch*, 20 August 1887. Photo courtesy of *Punch*

## 73 — 1887

## John Tenniel, The "Fire Fiend"

Captain (later Sir) Eyre Massey Shaw (1828–1908, aka 'The Skipper') was to become as famous a London fire chief as James Braidwood whom he succeeded in 1861. Shaw was from the Irish gentry, served in the British army and learnt his trade as a firefighter in Belfast. In 1866, the publicly funded Metropolitan Fire Brigade (MFB) replaced the LFEE, and was to continue until 1903. As Chief Officer of the MFB, Shaw proved to be a brave (he led from the front and was injured several times), dynamic if autocratic leader and innovator: he increased the number of men and stations, favoured the use of horse-drawn, steam, fire engines, telegraphic and telephonic communication systems, and improved life-saving by adding more street escape stations. One, undated head-and-shoulders photograph of Shaw wearing the ornate metal helmet (silver for him, brass for his men) and the navy-blue uniform he introduced in 1866, shows that he had fair hair and a distinctive, straggly moustache and goatee beard. It seems that he was a strict disciplinarian with a rather brusque manner.

Although Shaw worked hard, he enjoyed the social whirl: he travelled around London in a private carriage driven by a coachman, belonged to gentlemen's clubs and moved in the highest social circles. For instance, he was a friend of 'Bertie', the Prince of Wales, later Edward VII, a royal who took a keen interest in firefighting and liked to attend fires attired in a uniform kept ready for him at the Chandos Street station, Charing Cross.

Aristocrats such as the Duke of Sutherland and the Earl of Caithness were also fire buffs and volunteers. (This indicates that the popularity of firemen extended across the whole of British society.) As Robyn Cooper has remarked: 'The romance and drama of firefighting was a class transcendent activity that could be enjoyed by middle- and upper-class volunteers [even though] the working-class (poorly paid) fireman, unlike the volunteers, was subject to the strictest discipline as well as long and boring hours of waiting for a call.' Shaw's critics, who included Americans and socialists, accused him of being a snob and a flunkey because of the favours he granted his upper-class friends. No doubt, Shaw thought he and the London fire service needed influential allies.

On 3 June 1871, Shaw the elegant man-about-town was depicted in a chromolithographic caricature by 'Ape' (that is, Carlo Pellegrini, born Italy in 1838, who came to England in 1865 and died in 1889), which appeared in *Vanity Fair* (1868–1914), a magazine that regularly published full-page caricatures of prominent Victorians. In Ape's full-length portrait, Shaw is dressed like a dandy in a top hat, frock coat, plus carnation and walking stick.

A bronze statuette of Shaw in uniform with his helmet by his side was sculpted by Count Gleichen (aka Prince Victor of Hohenlohe-Langenburg, 1833–91), a relative of Queen Victoria who was an admiral, a sculptor and an amateur fireman. The sculpture, entitled *The Extinguisher from some of his Chums*, was displayed in Shaw's drawing room and is now in London's Fire Brigade Museum. An oil rendition of Shaw in his uniform by portrait painter Henry Weigall Jr (1829–1925) was exhibited at the Royal Academy circa 1871.

So well known did Shaw become that Gilbert and Sullivan immortalised him in the song 'Oh, Foolish Fay' in the comic operetta *Iolanthe*. Shaw was in the audience when it premiered in 1882. He had a professional interest in London's theatres because he surveyed them and found that many were firetraps. His critical report prompted some improvements. A cartoon drawing by W. Fletcher Thomas appeared in *Ally Sloper's Half-Holiday* (a Victorian comic published from 1884 to 1923) in 1888. It featured a character created by Charles Henry Ross in 1867 called Alexander 'Ally' Sloper, F.O.M. (Friend of Man, a drunk with a huge red nose who slopes off down alleys to avoid paying rent) dressed as a fireman holding a hose which squirts water into Shaw's face and knocks him off balance. Sloper has established a commission to inquire into the safety of theatres against fire and is seeking Shaw's assistance.

Like Braidwood before him, Shaw distilled his knowledge into a manual entitled *Fire Protection* published in 1876. In 1891, after disagreements with his new employers – the London County Council – Shaw resigned his post and was knighted by Queen Victoria on his final day of service.

In the cartoon illustrated, a wood engraving published in the humorous magazine *Punch* in 1887, Shaw is shown as a man of action: he is dressed for firefighting – uniform, helmet, jackboots – holds a branch hosepipe in his left hand and a short, spiked/wedge axe in his right, and is subduing a scaly monster on the ground. The latter – 'The Fire Fiend' – wears a helmet-cum-fire-grate that is aflame, holds a lighted torch in his right hand and personifies the negative aspect of fire. (Another personification often used by illustrators is a red devil.) During the nineteenth-century, as Robyn Cooper has remarked, ideals of chivalry dating from the Middle Ages were incorporated into Victorian conceptions of the gentleman. This cartoon, for instance, echoes images of St George defeating a dragon.

John Tenniel (1820–1914), a prolific illustrator best known for his Alice in Wonderland images, was the artist responsible. The caption reads: 'Shaw the lifeguard man. "I've got him under: – but I must have support to *keep him down!*"'

Evidently, the cartoon was an instance of visual propaganda on Shaw's behalf, to assist his constant struggle to obtain more resources for the MFB in order to improve and expand it. Later, when Shaw resigned after 30 years of service, *Punch* lamented his departure by publishing another cartoon showing him removing his helmet and belt in front of a fire engine. Beneath the cartoon, entitled *The Fire King's Abdication*, was a laudatory poem.

Around 1890, an unknown studio photographer took a half-length portrait of Shaw. Standing to attention against a plain background, he gazes directly at the camera (it seems he was a tall man with steely blue eyes). He is dressed in his uniform with its gleaming epaulettes, buttons, belt buckle and medals, and he holds a polished silver helmet in his right hand. The photograph has the gravitas and formality of an official representation.

When Shaw died aged 80 in August 1908, his funeral was naturally attended by many firemen. He was buried in a modest grave near Fireman's Corner in Highgate Cemetery, north London. A fireboat commissioned in 1935 was named *Massey Shaw* after him. In 1940, it assisted the evacuation of British troops from Dunkirk. Rudolf Haybrook, a fireman-artist, was on board and made a pictorial record of what he witnessed. In 1984, a full-length biography of Shaw by Ronald Cox was published, which concludes with the claim: 'Captain Shaw was the greatest fireman London – and perhaps the world – has ever seen.'

74 Emile Renard, *Death of Lieutenant-Colonel Froidevaux of the Sapeurs Pompiers of Paris*, (1887). Painting, oil on canvas, 68 x 93 cms. Photo courtesy of Musée Crozatier, Le Puy-en-Velay. © 1986 Direction des Musées de France

## 74 — 1887

### Emile Renard, Death of Lieutenant-Colonel Froidevaux of the Sapeurs Pompiers of Paris

## 75 — 1887

### Emile Renard, Death of Lieutenant-Colonel Froidevaux of the Sapeurs Pompiers of Paris

It is evident from this painting by Emile Renard and similar works by artists such as Édouard Detaille and Eugénio Alvarez-Dumon that some French artists active in the second half of the nineteenth century were willing to follow Courbet's earlier example in choosing firefighting as a subject. Renard, who lived from 1850 to 1930, was an academic artist who produced portraits and genre scenes but is now little known. His canvas depicts the death of a Parisian fire chief and ex-military officer – Lieutenant-Colonel François Xavier-Eugène Froidevaux (1827–82) – while on duty. In the centre of this animated composition, several firemen cluster around his limp helmetless body rendering assistance and dragging beams away while behind them, on the right, a fireman battles flames and smoke with a hosepipe. In the background on the left, reinforcements – fire companies and steam engines – hurry to the fire ground. Most of the painting is dark but dawn appears to be breaking behind buildings in the distance. The scene is illuminated by a lurid orange light from the fire on the right of the picture, which glitters on the metal helmets of the rescuers. Since the painting is executed with broad brushstrokes, it was presumably a preliminary study for a more polished, final version. The latter – which differs considerably from the former – is now in the collection of a Parisian firefighting museum and illustrated in Raymond Deroo's history of French fire brigades.

Froidevaux died on the night of the 6 and 7 of October 1882 while fighting a serious conflagration at Bois & Company, a metal utensils factory and store in the Boulevard de Charonne, which contained inflammable materials such as wood and varnish. While

Froidevaux was directing his men, walls collapsed and a falling beam smashed his skull. The Minister of the Interior & War attended his funeral and later, he was honoured by having a street in Paris named after him.

## 76 1887

## Unknown photographer, Officers of the Girton College Fire Brigade

76 Unknown photographer, *Officers of the Girton College Fire Brigade*, (1887). Photo courtesy of the Archive of Girton College Library, Cambridge

In Britain, volunteer fire brigades based in the suburbs and the countryside continued for some time after municipal professional services were established in the centres of major cities. In addition, there were numerous small brigades consisting of part-timers, operated by private companies and institutions such as collieries, country houses, factories, department stores, hotels, hospitals, nursing homes, schools and colleges of education.

Since residential colleges such as the Royal Holloway College, Egham, Surrey (founded in 1886) and Girton College, near Cambridge (founded in 1869) were women-only, their fire brigades were too. Sarah Marks (1854–1923, aka Hertha Ayrton), who was to become a noted scientist and inventor, and Gertrude Jackson, established Girton's brigade in 1879. (The College was a fire risk because it was illuminated at the time by oil lamps and candles.) Its members received tuition in the rudiments of firefighting from Captain Shaw. They drilled regularly and mastered the arts of climbing ladders, using warning rattles, ropes and canvas buckets, operating a small cylindrical hand pump (a 'corridor' pump mounted on two wheels made by Shand Mason), throwing glass grenade extinguishers, unrolling hoses, lowering bodies from the upper storeys of buildings and treating burns.

75 Emile Renard, *Death of Lieutenant-Colonel Froidevaux of the Sapeurs Pompiers of Paris*, (1887). Painting, oil on canvas. Paris: Musée de Traditions de la Brigade de Sapeurs-Pompiers de Paris. Photo courtesy of Sapeurs-Pompier de Paris

A group photograph dated 1887 shows eleven members of the brigade dressed in plain dark outfits with long full skirts posing with their equipment. The brigade even had its own song and fancy dress costumes in the shapes of buckets and hoses. Fortunately, Girton's Fire Brigade, which continued until 1932, was not required to extinguish any fires at the College. However, in 1918, it did turn out when a nearby haystack caught alight.

At the Royal Holloway College, whose buildings were encircled by fire hydrants, a brigade of around fifty was formed and adopted a uniform of blouses, blue serge suits and sailor hats. Circa 1898, photographs were published in the magazine *The Girl's Realm* (1895–1915) of brigade members engaged in stirrup pump and hosepipe drills. Two of the tallest and strongest students were assigned to direct the hose nozzle and they found the sudden rush of water through it resembled 'an electric shock'. Unlike their counterparts at Girton, the Royal Holloway brigade did have to extinguish a real blaze when a room hung with flimsy draperies caught fire.

In another photograph of 'lady' firefighters, taken during the 1880s, the scene is rather artificial in terms of the setting – the forecourt of a firehouse with exotic plants in the background – and the static poses adopted by the female firefighters. Although equipped with a steam pump, axes, helmets and hosepipes, the women seem too clean and smartly dressed to be serious about their task. One suspects this photo records a publicity event marketing the equipment on display. Further images of women dressed in firefighting uniforms were to appear during the early twentieth century when British suffragettes mounted their 'Votes of Women' campaign.

As Robyn Cooper has noted, the advent of female firefighters met with a hostile reception from the press: 'The *New York Times* in a mocking report [1884] condemned the venture on the grounds of immodesty. "That firegirls should actually ascend ladders in full gaze of the public, and while the fierce light of fire plays about their ankles, is of course, unthinkable." ' The second reason given by the *New York Times* was that the duties of a fireman and a lady were entirely incompatible. In seeking to be 'firegirls' these women invaded and threatened the male sphere and violated their own femininity.

## 77 1887

## Unknown illustrator, The Red Caps

77 Unknown illustrator, *The Red Caps*, (1887). Print published on the cover of *The Boys of New York: A Paper for Young Americans*, Vol. 13, No. 628, 27 August 1887. Photo courtesy of the Dime Novels & Penny Dreadful Collection, Stanford University Library

As literacy expanded and printing technologies and distribution systems improved during the nineteenth and early twentieth centuries, increasing numbers of books, newspapers and magazines were produced for the mass market. Among them were popular weekly magazines called 'dime novels' in the United States and 'penny bloods' and 'penny dreadfuls' in Britain. Such descriptions emphasize how cheap they were. Many were aimed at children and teenagers, especially boys. American titles included: *All Around Weekly*, *Brave and Bold Weekly*, *Boy's Weekly*, *The Boys of New York*, *Happy Days*, *The New Sensation*, *Pluck and Luck*, *Saturday Night*, *Secret Service*, *Tip Top Weekly*, *Wide Awake Weekly*, *Work and Win* and *Young Men of America*. British titles included: *Boys Own Paper*, *The Boys of England* and *True Blue Library*. Initially, the magazines carried black-and-white illustrations on their covers, but by the early 1900s, the illustrations were in full colour.

The lurid melodramatic tales the magazines contained featured famous murderers, highwaymen, outlaws and pirates but also hero-figures such as detectives, secret agents, Wild West frontiersmen and firemen. The latter included such characters as Oscar Everding, Fred Fearnot and Ronald Dare, and boy-heroes with whom juvenile readers could identify: Young Wide Awake, Chief of the Boy's Brigade, Phil the Boy Fireman and Plucky Boy Fireman.

As one might expect, the 'saved' motif – males rescuing swooning females from house fires – or the 'speeding fire engine' motif were frequently employed. While many stories were fictional, some were based on real events. As late as 1908, writers were recounting the exploits of New York's nineteenth-century volunteer companies.

On the cover of *The Boys of New York: A Paper for Young Americans* (published by Frank Tousey in New York from 1886 to 1894) for August 27th 1887, a black-and-white illustration by an uncredited artist introduces a story entitled 'The Red Caps' or 'The Fire-Boys of Boylston' by Robert Lennox, an author who penned several other firefighting stories. In this instance, a fire-boy wearing a man's hat is rescuing a horse from a burning building. A cloth has been wrapped around the horse's head to reduce its terror and the fire-boy rides his mount at full pelt through an open doorway. Rather absurdly, it looks as if the horse and rider are leaping into space high above the ground. It seems unlikely that either would have survived the fall. Inside a circular vignette is a view of a firehouse. The story may have been based on a true event because Boylston is a town in Massachusetts and there is a Boylston Street in Boston.

Clearly, illustrated dime novels – a large collection of which is preserved by Stanford University Library – helped to popularise the heroic image of the firefighter among American and British youth.

## 78a/b 1887

### Charles A. Orleans and Alexander Doyle, The Firemen's Monument

In New Orleans there is a well known cemetery – Greenwood, established on swampland in 1852 – which has an imposing memorial erected in 1887 to mark the 50th anniversary of the founding of the Fireman's Charitable & Benevolent Association. The latter was a mutual aid society (incorporated in 1835) of the volunteer firemen who protected the city before the New Orleans Fire Department, with its full-time paid profes-

78a and 78b Charles A. Orleans and Alexander Doyle, *The Firemen's Monument*, (1887). Memorial, marble and granite. New Orleans: Greenwood Cemetery. Photos courtesy of The Historic New Orleans Collection, New Orleans sionals, was established in 1891. The Monument was also intended to honour the memory of volunteers who had sacrificed their lives during the previous half century.

*The Firemen's Monument* stands on a low earth mound and consists of a 13.8 metre-high architectural canopy – four arches surmounted by a steeple – made from Hallowell Maine granite, which shelters a life-size, Italian marble statue of a fireman dressed in uniform and helmet who is holding the nozzle of a hosepipe. The names of 23 volunteer companies are inscribed on the base. The structure's form and its neo-gothic style were inspired by the Sir Walter Scott Monument (1844) in Edinburgh, which is a combination of architecture and sculpture by George Kemp and John Steell. The New Orleans Monument was designed and erected by the builder Charles A. Orleans at a cost of $14,000, while the artist Alexander Doyle conceived the figure and the Carlo Nicoli Studio of Carrara (founded in 1863) carved it.

Doyle (1857–1922), who was born in Steubenville, Ohio, spent a dozen years studying sculpture in Florence and Rome before establishing a studio in New York. He was a successful academic-style sculptor who generated a series of marble and bronze statues of American presidents and Civil War heroes for clients across the United States. His rather stolid figure of the fireman faces south east towards the head of Canal Street, New Orleans' main thoroughfare. The single figure, of course, is an ideal-type intended to stand in for all New Orleans' volunteer firefighters.

79 Charles Vigor, *Saved*, (c. 1890). Painting, oil on canvas, 195.6 x 134.5 cm, Moreton in Marsh, Gloucestershire: The Fire Service College Collection. Photo courtesy of Michael Kernan.

## 79 — 1890

## Charles Vigor, Saved

## 80 — 1913

## Unknown photographer, A Smoko Night in a Melbourne Fire Station with MFB banner in background

Little is known about the painter Charles Vigor, apart from the fact that he was a Victorian academic artist active in London between 1881 and 1904, who painted portraits and figurative subjects of a sentimental kind. *Saved*, it seems likely, was inspired by earlier works such as Millais' *The Rescue* and Louis Maurer's *Prompt to the Rescue*. Unlike Millais' fireman, who is seen from the side, Vigor's is seen from the front and he cradles only one child – a pretty, young, unconscious female – in his arms, upon whom he gazes with concern.

The fireman emerges from the doorway of a middle-class house from which flames and smoke are pouring and descends debris-strewn steps across which lies a hosepipe. Some descriptions of the painting claim that the house was the Georgian one – Winchester House – used by Captain Shaw and his family at the Southwark headquarters of the MFB. The fireman wears a brass helmet – introduced by Captain Shaw in the 1860s – rather than a leather one.

*Saved* was exhibited at the Royal Academy in 1892 and later bought by a Southport Justice of the Peace named James Wood. Appropriately, the painting is now owned

80 Unknown photographer, *A Smoko Night in a Melbourne Fire Station with MFB banner in background*, (1913). Photo courtesy of The Fire Services Museum of Victoria, Melbourne

81 Jay Hambidge, *The Death of Chief Bresnan*, (1898). Magazine illustration, 12.5 x 19 cms. Published in *The Century Magazine*, February 1898, p. 487. Author's collection

by Britain's Fire Service College (a training school), Moreton-in-Marsh, Gloucestershire. A chromolithographic print (credited to 'Colourviews') was based on the painting. I say 'based', because the print is considerably lighter than the painting and has differences of detail. At one time, the print was displayed in London's fire stations. (Currently, The Fire Service College markets it.) In 1896, the print was used to adorn calendars such as that issued by Stephen Hosken, a family grocer and merchant located in Copperhouse, Hayle, Cornwall. Subsequently, *Saved* became known globally; for example, it was reproduced on Russian and Eastern European postcards and an enlarged copy of it, made by a Melbourne fireman, appeared on a banner carried in Labour Day marches by Australian firemen belonging to the United Firefighters Union. The banner was also displayed during social gatherings called 'smoko nights' held by the Metropolitan Fire Brigade of Melbourne at its headquarters fire station. A photograph dated April 1913 depicting one of these gatherings is preserved in the Fire Services Museum of Victoria (which now occupies the headquarters building).

Furthermore, Vigor's image is still in vogue. For example, it is the source for a popular ornament: a Royal Worcester Porcelain figurine (24–cms-high) and for an image on a hand-painted enamel Crummel Pill Box marketed by Phoenix World Reproductions, Gravesend, Kent. This company also manufactures a range of firefighter collectibles in the form of bronze figurines. *Saved* has also been reproduced on the cover of a magazine entitled '*Saved*' issued by the Friends of The Fire Services National Museum Trust.

## 81 — 1898

## Jay Hambidge, The Death of Chief Bresnan

## 82 — 1887

## Unknown illustrator, Portrait of John J. Bresnan

Edward John 'Jay' Hambidge (1867–1924) was an American figure, landscape and genre painter whose work was also reproduced as illustrations in popular magazines. For instance, several of his paintings were used to illustrate Jacob A. Riis' 14–page article 'Heroes who fight fire' published in *The Century Magazine* (Vol. 55, No. 4, February 1898). Riis (1849–1914) was born in

82 Unknown illustrator, *Portrait of John J. Bresnan*, (n.d.). Engraving in A. Costello's book *Our Firemen*, (1887)

83 John G. Segesman, *Volunteer Firemen Monument*, (1896/1905). Memorial/public sculpture, bronze and granite, c. 12 m high. Austin, Texas: State Capitol Grounds. Detail from C00324–A, Austin History Center, Austin Public Library

Ribe, Denmark, trained as a carpenter and then, in 1870, he moved to the United States. Today, he is primarily remembered for his socially concerned documentary photographs of the vile conditions in the New York slums taken during the 1880s and 1890s. (Riis was a social reformer who identified with the poor and exploited; consequently, it was logical that he should also have wanted to honour public servants.) However, as this article demonstrates, he was a significant writer/journalist too. In his article, Riis claims to have known some of New York's firemen for many years and so he had a personal motive for wanting to hail them as heroes and to recount some of their daring rescues and the sacrifices they had made for the public.

One such sacrifice was made by John J. Bresnan, Chief of the Sixth Battalion with headquarters in Hook & Ladder Company No. 3, who died on a snowy 29 December 1894 while leading his men tackling a fire in Cassidy & Son's factory on West 24th Street. Like so many American firemen, he was of Irish extraction and contemporaries reported that he knew fire 'like a botanist knows a flower', that he was 'a quick and utterly fearless leader' and modest about his life-saving exploits. Assistant Foreman John L. Rooney, who had been awarded a medal in 1883 for saving the life of Miss Ida Small a year earlier, was also killed. A water tower on the roof whose wooden supports had burned away fell and crushed the two men.

The composition of Hambidge's illustration is unusual because one quarter is light and the other three quarters are dark. (Hambidge was also an art instructor and theorist, and so his composition may have been influenced by his ideas of 'dynamic symmetry' which he promoted in a series of publications beginning in 1919.) The darkness serves two purposes: it conveys the conditions of low visibility in which firefighters so often have to work and symbolises the presence of death. In the centre, five helmeted firemen climb across a pile of debris inside a building; two are silhouetted against light penetrating the gloom from above. Water is cascading down. The leading fireman wields an axe while the second uses a hook to test a wooden beam. In the bottom, right-hand corner, there is a discarded fire helmet and underneath debris can be seen the corpse of its owner.

Chief Bresnan's three children became orphans. However, undeterred by Bresnan's fate, his male descendents continued the family tradition of firefighting: his son, grandson and great-grandsons all joined New York's Fire Department.

## 83 — 1896/1905

### John G. Segesman, Volunteer Firemen Monument

During the 1880s, the State Firemen's Association of Texas obtained permission to erect a granite monument to fallen, volunteer firemen in the grounds of the new State Capitol building in Austin, Texas. Completion, however, did not occur until 1896. A firm called Jaeggli & Martin of Brenham was contracted to build the Monument and Frank Teich (1856–1939), a sculptor and stonecutter from San Antonio, was commissioned to design it and to cut blocks of grey granite from a local quarry.

Like so many sculptors in nineteenth-century America, Teich was originally an immigrant. He was born in Lobenstein, Germany, studied at the University of Nuremberg, and then served as an apprentice to the sculptor Johannes Schilling. He arrived in the United States in 1878 and after working in several states, he settled in Texas and opened a marble yard in San Antonio in 1885. During a busy career, he executed many monuments and public sculptures in granite, marble and bronze, and earned the title 'Father of the granite industry of Texas'.

Teich carved the figure of a fireman holding a hosepipe and this was placed on a tall tapering obelisk which itself rested on a pedestal and base. Old photographs reveal that the Monument was about 18 metres high and occupied a prominent position on an open slope directly in front of the huge Capitol building designed by the architect Elijah Myers. The Monument included inscriptions of the names of the fallen and was embellished with firefighting emblems. With much pomp and ceremony, the Monument was unveiled at a statewide meeting of the Firemen's Association on 7 July 1896. Over 100 fire companies attended, and music and songs were provided by a brass band accompanied by a chorus of 600 voices.

However, due to structural problems and 'public disapproval', within a few years, the obelisk was removed and the Teich sculpture was replaced by a figure modelled by John G. Segesman and cast in bronze by the W.H. Mullins Art Metal Works of Salem, Ohio. (The second version of the Monument is much reduced in height.) Another dedication ceremony was considered necessary and took place in 12 May 1905. Today, the Monument is located just inside the entrance gates to the Capitol grounds, which are now park-like.

Like Teich, Segesman was an immigrant. He was born near Berne, Switzerland in 1865 and moved to the United States at the age of 16. During the 1890s, he made his home in Salem and learnt to model the human figure in the workshops of W.H. Mullins, a company that made huge statues from bronze and sheet copper for clients around the world. Mullins had originally planned to send Segesman to art school but he never went because the firm was always too busy. Thus, Segesman was largely self-taught. Once promoted to the rank of sculptor, he produced a series of public sculptures: figures personifying justice, light, progress and victory, statues of politicians and generals on horseback, busts of presidents, bas-reliefs of Civil War battles, sculptures of animals such as moose and elk, etc. Segesman retired from Mullins in the 1920s and died in 1953 aged 88.

Segesman's bronze fireman is exceptionally slim and tall probably to compensate for the foreshortening that would occur when the figure was viewed from below. His moustachioed figure is smartly dressed in a double-breasted, buttoned-up overcoat and he poses in an extremely dignified and straight-backed fashion. In his left hand, he holds a small child dressed in nightclothes who is awake and pointing, and in his right, he carries an oil lamp. A large helmet dwarfs his head. This figure, which conveys an impression of calm deliberation (the sign of an experienced unflustered firefighter), is one of most artistically accomplished and impressive of American firefighter statues.

This was not Segesman's only statue of a fireman. Two years earlier, in 1903, he had sculpted another, very similar bronze or sheet copper figure of a fireman with child and lamp for the Fireman's Association of the State of New York. It was dedicated to the old Elmira Exempt Fire Association and placed on a granite base over twelve graves belonging to Elmira fallen firemen in Woodlawn Cemetery, Elmira, New York. The main difference between the two figures is that the Austin one holds his lantern down by his side whereas the Elmira one holds his aloft. Viewers familiar with the work of the British Pre-Raphaelites will be reminded of William Holman Hunt's painting *The Light of the World* (1851–53), which depicts Christ carrying a lamp.

## 84 **1897**

## Julio Martínez Zapata and Don Augustín Querol y Subirats, Monumento a los Bomberos

One of the largest and most elaborate nineteenth-century tombs/memorials to volunteer firefighters was the *Monumento a los Bomberos*, erected in the Necropolis Christóbal Colón cemetery, Havana, Cuba in 1897. The cemetery, which contains many statues, ornate tombs and mausoleums, was established in the 1860s and 1870s. It was named after Christopher Columbus and is

84 Julio Martínez Zapata and Don Augustín Querol y Subirats, *Monumento a los Bomberos*, (1897). Marble and granite tomb/monument, Necropolis Christóbal Colón cemetery, Havana, Cuba. Photo: Stew Thornley

now a popular tourist attraction. Postcards for tourists have been produced with photographs of the Firemen's Monument on them.

The Monument was a tribute by the people of Havana to 28 firemen who lost their lives on 17 May 1890 while tackling a blaze in Mercaderes Street. It is made from marble and granite, and features a balustrade with steps leading to a large base on which sit allegorical female figures. From this rises a richly carved column on top of which there is a Christian cross and the figure of an angel gesturing towards heaven with one arm while holding the corpse of a fallen fireman with the other. Two Spaniards were responsible for the design: the architect Julio Martínez Zapata and the sculptor Don Augustín Querol y Subirats.

## 85a 1897

**Giuseppe Moretti, Memorial to the 73rd New York Infantry Volunteers**

## 85b 1897

**detail**

## 86 1924

**Giuseppe Moretti, Fireman's Monument**

85a Giuseppe Moretti, *Memorial to the 73rd New York Infantry Volunteers*, (1897). Memorial/public sculpture, bronze figures and granite base. Gettysburg: Excelsior Field. Photo courtesy of Kathleen Herzog, sagbottom.com

The Battle of Gettysburg, Pennsylvania, which occurred in July 1863 and lasted three days, was one of the fiercest of the American Civil War and resulted in great loss of life on both sides. The Union Army of the Potomac was eventually victorious. Today, the site of the Battle is a National Cemetery and Military Park, which has many monuments. One of the oddest – because it features an over life-size bronze statue of a fireman wearing a helmet and holding a trumpet standing next to a statue of a soldier with a rifle – commemorates the 73rd New York Infantry Volunteers (aka 2nd Fire Zouaves and 4th Regiment Excelsior Brigade). The reason for the fireman statue is that this regiment (at first 814 strong) primarily consisted of firemen and runners from Manhattan who had volunteered to fight for the Union in May 1861. (Clearly, the two bronze figures standing shoulder-to-shoulder illustrate their dual identity.) On the afternoon of 2 July 1863, the New Yorkers fought a battle in a Peach Orchard. Out of a force of 349 men commanded by Major Michael W. Burns, they lost 51 killed and over 100 wounded and missing.

The Memorial was commissioned and funded by the State of New York and by the City's fire companies. It was dedicated during September 1897. A New York architectural firm – Hoffmann & Prochazka – was responsible for the overall design and the Gorham Manufacturing Company undertook the casting. The sculptor selected to model the figures was Giuseppe Moretti (1857–1935), one of the most famous, prolific, academic artists of the day. During the course of a long, commercially successful career, he ignored the challenge posed by modern art.

85b detail

86 Giuseppe Moretti, *Fireman's Monument*, (1924). Memorial/public sculpture, bronze, 2.4 m high. Harrisburg, PA: Riverfront. Photo Jeb Stuart courtesy of Tina King

Moretti was born in Siena, Italy into a family of art patrons and later moved to Florence to become an apprentice to the sculptor Giovanni Dupre. Moretti then developed his carving skills in the marble quarries of Carrara. He went to the United States in 1888, established studios in New York and Pittsburgh, plus foundries and marble works, and undertook commissions for public sculptures and monuments across the country. His most famous piece was Vulcan, a giant cast-iron statue, executed in 1904. He lived in the United States for 40 years but only became an American citizen in 1923; he died in Italy after retiring there.

In 1924, Moretti created a second statue honouring firemen who became soldiers: it was a 2.4–metre-high bronze figure of a somewhat bland, idealised fireman-soldier clasping a child under his cape and holding a sword wreathed with a symbol of victory. It paid tribute to the firemen of Harrisburg, Pennsylvania who had fought in World War I. Decorating the statue's granite pedestal was a bas-relief showing firemen in action. Recently, this sculpture – which is located in a pleasant riverside setting – has been cleaned and restored. Moretti said of it: 'In my subject ... I have typified the two great virtues lying nearest the heart of our nation – the preservation of liberty and salvation from destruction by fire ... Artists have illustrated the picturesque dash of the fire horses, the thrilling runs to the fire. In my memorial, I have shown the noble souls of the firemen and what they mean to our nation.'

During the autumn of 2002, Moretti's artistic achievements were celebrated in an exhibition, curated by Miriam Rogers Fowler, held at the Birmingham Museum of Art, Alabama. (Moretti had lived in Alabama for many years in order to exploit its deposits of white marble.) Also in 2002, the artist Gary Casteel and the Four Winds Studio and Gallery, Gettysburg, which offers for sale a Gettysburg Battlefield Monument Collection consisting of small replicas, announced the addition of the 73rd Monument, 38 cms high, in two editions: a limited one of 100 made from Cold Cast Bronze and an unlimited one made from Pecan.

87 W.L. Cottrell, *Firemen's Monument*, (1898). Memorial/public sculpture, life-size bronze. St Joseph, MI: Lake Bluff Park. Photos courtesy of the Krasl Art Center. Left, before restoration; right, after restoration

## 87 — 1898

### W.L. Cottrell, Firemen's Monument

## 88 — 1900s

### Unknown photographer, Firemen's Monument, St Joseph, Mich. Erected in Honor of Firemen who died in Performance of Duty

This bronze figure of a fireman with a finely modelled face who is carrying to safety a young girl wearing a nightdress dates from 1898. W. Liance Cottrell, a New York architect, has been credited with the design of the Monument and the Harrison Granite Company fabricated the tiered granite base decorated with bronze wreaths and firefighting insignia. Fireman Fred Alden and five-year-old Mabel Synder posed for the figures. This Monument was erected by the local Volunteer Fire

88 Unknown photographer, *Firemen's Monument, St Joseph, Mich. Erected in Honor of Firemen who died in Performance of Duty*, (early 1900s). Colour picture postcard, 14 x 9 cms. Published by Burkhard Bros., St Joseph, Michigan. Posted 3 July 1911. Author's collection

Department at a cost of $2200 to commemorate five of their colleagues killed during the burning of Yore's Opera House, Benton Harbor, Michigan in September 1896. It is flanked by two lamps and was one of the first in the United States to be illuminated at night via self-contained electrical fixtures. The Memorial was dedicated on 31 May 1898 and re-dedicated on 20 May 2000 following a restoration of the statue by Venus Bronze Works of Detroit.

As mentioned earlier, picture postcards were one medium through which firemen's memorials were celebrated and became more widely known. Reproduced is such a card depicting the *Firemen's Monument* in Lake Bluff Park, which includes a female pedestrian and a horse-drawn carriage with three firemen aboard.

89 Peck & Snyder, New York, *Firemen's complete outfits* (1878). Advertisement published in a Peck & Snyder Sporting and Professional Goods Catalogue. Photo courtesy of the Rucker Archive

**89** — **1878**
**Peck & Snyder, New York, Firemen's complete outfits**

**90** — **1900**
**Cecil Alden, Cadbury's Cocoa**

**91** — **1888**
**Unknown photographer, London: Ready for a call**

**92** — **1900**
**Unknown artist, St Jacobs Oil**

**93** — **1907**
**Jules Abel Faivre, The Rescue**

**94** — **1930s**
**Unknown artist, We are having a hot time**

**95**
**J.B. Taylor, The fire is under control**

Images of firemen appeared in advertisements as early as 1878 – witness the picture of an American fireman wearing a red shirt and blue trousers in an advert for Peck & Snyder of New York, which manufactured firemen's clothing and equipment.

Companies selling fire extinguishers and insurance were naturally also keen to employ images of firemen in their advertisements.

By the end of the nineteenth century, firemen had become such popular figures that advertisers could use them as role models to market products unrelated to fire, such as warm drinks and liniments. For example, Cadbury's advert for Cocoa designed by Cecil Alden in

90 Cecil Alden, *Cadbury's Cocoa*, (1900). Advertisement. Photo courtesy of Cadbury Trebor Bassett, Bournville, Birmingham

91 Unknown photographer, *London: Ready for a call*, (1888). Photograph. Dundee: J. Valentine & Co. Photo courtesy of St Andrew's University Library, Photographic Collections

1900, which employed the slogan 'Makes strong men stronger' and which was published in magazines such as the *Illustrated London News*.

Rather oddly, the fireman sporting a handlebar moustache about to sip a saucer of hot cocoa is ignoring the fire raging behind him and turns his back on two colleagues who are tackling it. On the left, belching steam is a pump, which was probably manufactured by the London engineers Shand Mason & Co. The source of Alden's principal figure seems to have been a photograph taken in 1888 for J. Valentine Co. of Dundee, because it depicts a very similar fireman from the Metropolitan Fire Brigade posing in front of a Shand Mason steam engine.

In a print advert dating from around 1900 for St Jacobs Oil, the claim 'The Brave men of the Fire Brigade know the value of St Jacob's Oil for Rheumatism, Lumbago ...' appears next to the image of a handsome brass-helmeted fireman wearing a MFB uniform who is rescuing a swooning damsel in distress. (The patent medicine was originally German. August and Charles Vogeler – father and son – established a company in the 1870s in Baltimore to sell it; later, in 1900, a British firm bought the company.) To escape the flames and smoke of a house fire, he clambers over the iron railings of a balcony and prepares to use a rope to descend or swing to safety. (To modern eyes, this action-man image is highly reminiscent of jungle scenes in which Tarzan, clutching Jane, uses vines to swing from tree to tree. However, Edgar Rice Burroughs' first Tarzan story was not published until 1912.) It is perhaps significant

92 Unknown artist, *St Jacobs Oil*, (c. 1900). Advertisement. Courtesy of London Fire and Civil Defence Authority

93 Jules Abel Faivre, *The Rescue*, (c. 1907). Colour picture postcard, 9 x 14 cms. Author's collection

that when Captain Shaw was in charge of the MFB, he recruited ex-Navy seamen because he thought their skills – for example with rope and climbing rigging – made them ideal for firefighting.

As Robyn Cooper has pointed out, firefighters saved men too 'but in verbal and visual narratives the rescue of men is almost completely absent, for the obvious reason that such representations would place men in a dependent situation. Scenes of the rescue of women were an enhancement of the fireman's masculinity, a reassuring affirmation of gender difference, and a visible demonstration of man's power over women.'

The thought of virile firemen holding half-naked, semi-conscious maidens clearly had erotic connotations for many viewers, particularly French artists. This is evident in Eugénio Alvarez-Dumon's painting *Un Sauvetage à Paris* (1886, Musée Crozatier), which depicts a fireman in a dark bedroom rescuing a swooning female whose left breast is fully exposed. It is also evident in early twentieth-century comic illustrations by Jules Abel Faivre (1867–1945, best-known as a poster artist), which he created for books and magazines, showing firemen descending ladders with women in their arms or sitting on their shoulders. A saucy postcard was based on the latter image in which the fat woman straddling the fireman's shoulders rests her hands contentedly on his phallic-shaped helmet. An American postcard dating from the 1930s depicting a fireman on a ladder with a scantily-dressed female in his arms was accompanied by the suggestive, punning caption 'We are having a hot time'.

Flames and heat have long been associated with the 'burning' passion of love. The widely held opinion that firemen are lustful individuals has been made visually explicit in several comic postcards. For instance, a British one issued by Bamforth & Co. Ltd., probably in the 1930s, depicts a 'young ladies hostel' just after a fire has been extinguished. Through two open windows on the first floor firemen can be seen embracing partially-dressed young women while in the foreground a fire chief tells the hostel's female supervisor: 'The fire is under CONTROL now madam!' To which she replies sarcastically: 'Yes, chief, the FIRE is – but what about the

94 Unknown artist, *We are having a hot time*, (1930s?). Art-Colortone picture postcard, 9 x 14 cms. Chicago, IL: Curt Teich & Co., Inc. Card postmarked 1940. Author's collection

95 J.B. Taylor, *The fire is under control ...*, (n.d.). Colour picture postcard, 9 x 14 cms. Holmfirth, Yorkshire: Bamforth & Co. Ltd., Comic series no. 786. Author's collection

FIREMEN?' In a second comic postcard, a seated fireman has an attractive parlour maid on his lap, while she has grasped the prime tool of his trade – a hosepipe – and is squirting water into his mouth in order to cool his ardour. Predictably, the card's caption reads: 'Cooling the flame.'

A third humorous card from around 1944 depicts an American volunteer fireman about to enter a bedroom in which an alluring female dressed in underwear and high heels reclines on a bed. She is reading a book entitled *Flaming Love* and smoking a cigarette. At the window, the fireman grins, tips his fire hat in greeting and remarks: 'I'm just a volunteer – always glad to look up an old flame!' A hosepipe is draped over his shoulder, but since it is not charged with water, it is in a limp condition that does not bode well for any lovemaking.

As we shall discover later, firemen themselves have also been represented as sex objects in calendars and plays.

## 96 — 1900

## Paul Gustaf Fischer, The Fire Brigade turns out in Coal Square

Fischer, a minor but skilled, prolific and popular Danish painter, photographer, printmaker and sculptor, lived from 1860 to 1934 and specialised in portraits, landscapes and urban scenes. So frequently did he depict the people and streets of Denmark's capital that he became known as 'Copenhagen's painter'. Fischer came from a middle-class background and middle-class patrons bought his work. His mother was Danish but his father was of Polish-Jewish origin. The latter was a painter and a colour and varnish manufacturer, and Fischer learned his craft early and worked in his father's factory. He then trained at the Royal Academy of Art in Copenhagen from 1876 to 1878. Later he travelled to France and Italy, and his works were exhibited internationally: in Chicago, Munich, Paris and Stockholm.

His mature painting style can be characterised as academic-naturalist enlivened with qualities of brushwork and rendering of light derived from French impressionism. The 1877 painting *Paris Street, Rainy Day* by Gustave Caillebotte (1848–94) seems particularly relevant to Fischer's carefully composed painting *The Fire Brigade turns out in Coal Square* (that is, Kultorvet, Copenhagen). This work also recalls the detailed figurative canvases of the Victorian era documenting everyday scenes of British society such as the Post Office and Derby Day. Fischer took many photographs and used some of them as reference sources because photographs of firemen in poses exactly like those in the painting exist. Other photographs show a fire engine being stoked.

*The Fire Brigade* ... depicts 19 members of Company 113 of the city's fire department tackling a blaze in what appears to be a clothing store called Stromper located on a corner. Part of the store appears on the left-hand side of the picture. An escape ladder propped up against it is being climbed by a fireman (*brandvaesenman* in Danish) with a hosepipe over his shoulder. A colleague is about to follow. The men's black helmets and uniforms are similar to those used by German firefighters. The Company is equipped with polished brass steam engines with red-painted wheels, one of which, appropriately, given the name of the square, is being stoked with coal on the right-hand side of the canvas. (A separate painting of this steam engine exists, which was produced as a preliminary study. The engine, named 'Surtur' after a fire giant of Norse mythology, was acquired by the fire brigade in 1900 and is now preserved in the Danmarks Tekniske Museum, Helsingør.) On the left, stands an alert and handsome young fireman who grasps a metal turnkey while awaiting the order to charge a hosepipe with water from a supply situated

96 Paul Gustaf Fischer, *The Fire Brigade turns out in Coal Square*, (1900). Painting, oil on canvas, 90 x 120 cms. Private collection?

beneath a cobbled street. Other firemen are handling hoses while, in the middle distance, an officer gestures and gives instructions. The officer is placed at the exact centre of the canvas.

Smoke drifting from the fire towards the centre of the square creates a misty effect, which partly obscures the imposing buildings in the background. Reflections and puddles indicate that the pavement and cobbles are wet and slippery. Since a large crowd has gathered, it is clear Copenhagen's citizens are fascinated by the spectacle of firemen in action. Many watchers are holding umbrellas so it must be raining. Several prominent male citizens wearing top hats stand in front of the crowd. A male spectator, third from the left, is holding a camera case. This is a portrait of renowned Danish photographer Paul Elfelt who at the time had a studio at the top of the building behind him.

At first, Fischer's composition seems complex because there are several centres of interest at different distances within the pictorial space to engage the eyes and encourage them to explore. Soon, however, one realises that everything is organised around a curve that starts on the left, sweeps round into the centre before exiting behind the shop's corner. Although Fischer's firefighting scene is a busy one, it is not chaotic because the firemen are performing their various tasks calmly and efficiently. Today, prints of this documentary-style painting are available via the website Artexpression.com.

There are two books about Fischer written by Danish authors, which reveal that he painted other firefighting scenes set in the streets of Copenhagen; for example, *Nikolaj Sprøjten* (1882), *Ildos i Skindergade* (1896) and *Idlebrand i Ny Vestergade* (1924). None, however, is as ambitious and impressive as *The Fire Brigade turns out in Coal Square*.

## 97 — 1901–06

## Butcher & Sons, Our Firemen

Before cinema commenced in the 1890s, Victorian audiences enjoyed shows of sequences of coloured images, which were fixed on glass slides and then projected in magnified form on to screens by lanterns made from

97 W. Butcher & Sons, *Our Firemen*, (c. 1901–06). Magic Lantern Slides, No. 717 Primus Junior Lecturer Series. Eight slides in set, each 8.3 x 8.3 cms. Photo courtesy of Henc R.A. de Roo, private collector

brass and mahogany. The lanterns, equipped with lenses and powerful light sources such as oxy-hydrogen limelight, were the precursors of cinema projectors. So-called Bi-unial projectors had two optical systems so that special effects such as dissolves from one image to another could occur. 'Soundtracks', that is, prose and verse being read aloud, generally accompanied the visuals. (The Magic Lantern Society based in Britain has copies of the printed texts that were issued with the slide sets.) The invention of magic lanterns occurred much earlier – around 1660 – and has been credited to the Dutch scientist Christiaan Huygens (1629–95). British firms that manufactured slides included Carpenter & Westley, Newton & Co., W.C. Hughes, Simpkin, Marshall & Co, York & Son and Bamforth & Co. Images were transferred lithographically or hand-painted on to glass using transparent pigments and then issued in sets of eight or twelve.

Firefighting was just one of many subjects treated by the slide makers. For instance, Millikin & Lawley of London issued a set of twelve 'Lithographic Colored Magic Lantern Slides' entitled *Bob the Fireman, or Life in the Red Brigade*. In 1890, J. Theobald & Co of London distributed this set in the United States. The circular, rather crudely rendered images contained some familiar scenes: interior of a fire station; a ladder escape being raised and hosepipes being used; horse-drawn engines dashing through the streets; firemen climbing ladders and rescuing women and children. One image is a direct crib from Millais' painting *The Rescue* and so the slide set must date from after 1855. Remarkably, the character 'Bob the Fireman' is still alive because he – a magician dressed as a fire chief – can be hired from the clowns' agency Funtime Entertainments Unlimited.

Another firm, W. Butcher & Sons of Blackheath, London, sold lanternslide sets under the trade name Primus from 1870 to 1906. Number 717 of its Junior Lecturers' Series B sets, entitled *Our Firemen*, was sold in a cardboard box for two shillings and sixpence, and came with printed lecture notes. There are eight images (square but with rounded corners) representing the usual scenes: fireman up a ladder wielding an axe; horse-drawn engines travelling at speed; a steamer, ladder and hosepipes at a house fire; fireman rescuing a baby ... but there is also a river scene with a massive warehouse blaze (the reference was to the 1861 Tooley Street inferno at which Braidwood died). In one dramatic, luridly coloured scene, a number of hose teams direct their jets of water in all directions to subdue several fires occurring in a narrow street. As the accompanying notes reveal, this slide depicted a serious fire in London's Jewin Street on 19 November 1897, which consumed six or seven streets, 300 warehouses and the church of St Giles, Cripplegate. The names of those who painted these scenes are not known but the illustrators T.W. Lascelles and W.H. Overend probably made them after paintings. The imagery of *Our Firemen* is more exciting and energetic than that of *Bob the Fireman*. Both sets of slides were aimed at children and their printed commentaries included many facts about the fire brigade's equipment and methods.

Pioneers of the cinema, such as James Williamson of Hove, used to give lectures illustrated by lanternslides to such audiences as children attending Sunday schools and there is little doubt that this kind of public education and entertainment was an important step towards narrative motion pictures. The cinema, of course, gradually ended magic lantern shows except as historical curiosities although some slide sets were issued as late as 1931. An example of the latter was *Mickey Mouse in 'The Firefighters'*, an eight-slide set produced by Ensign Ltd,

98 James Williamson, *Fire!* (1901). Film still supplied by the British Film Institute, London

99 Edwin Stanton Porter, 16 stills of scenes from the *Life of an American Fireman*, (1902–03). Edison Company. Photo courtesy of the British Film Institute, London

London by arrangement with Walt Disney, which was probably based on the 1930, black-and-white, animated cartoon called *The Fire Fighters*. Disney also sold a short film entitled *The Fireman*, starring Mickey Mouse, as a toy for screening via a clockwork-powered 'Movie-Jecktor'.

## 98 — 1901

### James Williamson, Fire!

## 99 — 1902–03

### Edwin Stanton Porter, 16 stills of scenes from the Life of an American Fireman

Shortly after the invention of cinematography during the early 1890s, black-and-white silent films about fires and firefighting began to be made. Clearly, the film pioneers considered the subject matter dramatic and thought it would have a popular appeal. Inevitably, the filmmakers reprised the motifs and themes already established in nineteenth-century paintings, prints and lanternslide shows. Even the form of film – a sequence of images linked by a narrative theme – was anticipated by composite picture spreads found in the illustrated magazines of the nineteenth century.

In the United States during 1896, a spate of brief documentary films were made with titles such as *A Morning Alarm*, *The Burning Stable*, *Fighting the Fire*, *Fire Department New Haven, Connecticut*, *Fire Department, N.Y. City, 1st and 2nd Alarms*, and *Fire Engine at Work*. Soon, fire departments in cities across the country were vying with one another to be filmed. Favourite subjects were rescue drills and fire engine runs. In 1897, during the filming of *Atlantic City Fire Department*, a speeding engine veered off course and crashed into the Biograph camera that was recording it; the negative survived and showed the engine heading directly for the camera. Once the novelty of seeing movement on screen had worn off, the entertainment value of short, unedited documentaries declined. Evidently, longer films that told stories were needed. For these, staged reconstructions would be required.

In 1901, the Scottish-born cinematographer, actor, inventor, writer and producer James Williamson (1855–1933) directed *Fire!* (280 feet, Williamson Kinematograph Company), a short drama-documentary filmed in Hove, Sussex with the help of the local fire brigade. Before devoting himself to filmmaking, Williamson had owned pharmaceutical and photographic shops in the town and had been fond of delivering lectures illustrated by lanternslides. (Early silent film screenings were often accompanied by explanatory lectures.) Before 1900, he produced over 100 one-minute trick films, comedies and actualities. In *Fire!* the plot has a beginning, middle and end, which is recounted via five scenes: 1) a policeman discovers a fire; 2) firemen respond to his alarm and hitch horses to three engines; 3) they rush down a street towards a house fire (a 'race against time', head on shot); 4) in an interior bedroom scene a man awakes to discover it is ablaze but he is rescued when a fireman enters via the window and hoses down the flames; 5) in an exterior scene the fireman carries the man down a ladder while another breaks into the house and re-emerges with a child in his arms; finally, a third victim jumps from upstairs and is saved by firemen holding a net. Williamson often projected his films to Sunday schools, so children must have enjoyed the sight of professional fire crews in action. *Fire!* has been praised for its innovative form as well as its lively content. Alice Liddel, for example, regards the film as 'a milestone' for using an 'edited series of interlocking scenes to create a narrative'.

One of the first filmmakers in the United States to tackle the subject of firefighting was William K.L. Dickson (1860–1935), an ex-assistant to the famous inventor Thomas A. Edison, a founding father of cinema. Dickson directed two of the documentary shorts of 1896 cited above. Another pioneer was Edwin Stanton Porter (1870–1941) who became, in 1901, principal cinematographer for the Edison Kinematograph Department founded eight years earlier. Porter had personal experience of the devastation that fire can cause when one occurred in his New York workshop. For his 1902–03 silent film *Life [or Story] of an American Fireman* (the title recalls the Currier & Ives print series of the 1850s and 1860s), Porter assembled nine separate camera shots filmed in New Jersey with the participation of the local fire department into a film with seven scenes, which lasted six minutes. Documentary footage was combined with re-enacted, dramatised scenes, which Porter organised. He was probably influenced by Williamson's *Fire!* because his plot is similar. Linked by dissolves, the scenes are these: 1) a fire chief inside a firehouse has a dream about a mother and child and senses danger; 2) an electric street fire alarm is activated; 3) firemen, aroused from slumber by the alarm, dress hurriedly and slide down poles; 4) they hitch up horses to appliances; 5) an exterior shot of engines emerging and driving off; 6) a pan or panorama shot depicts horse-drawn pumpers rushing down a street; 7) interior and exterior, a fireman – played by English-born actor Arthur White (1882–1957) – rescues a mother from a burning house and then returns for her child. The storyline, therefore, was formulaic. Like Williamson, Porter has been hailed as a pioneer of editing techniques and the narrative feature film. Today, he is chiefly remembered for his 1903 film *The Great Train Robbery*.

Both Williamson's and Porter's films can be viewed on video anthologies devoted to early cinema while some of Edison's can be viewed via the Internet and the Library of Congress's Motion Pictures collection.

## 100 — 1902

### Matthew Hastings, Captain Thomas Targee

This dramatic full-length portrait of Thomas B. Targee, an American fire chief who was probably of Irish Catholic origin, is really a history painting because its subject had been dead for 53 years when the work was completed. The artist, Matthew or 'Mat' Hastings (1834–1919), was a St Louis-based magazine illustrator and painter of portraits, historical subjects, religious murals, river scenes and landscapes. His parents came to the United States from Ireland in 1832 and he was born in Georgetown, D.C., and moved to St Louis, then 'Gateway to the West', in 1840 where he later became acquainted with the work of such American frontier artists as George Caleb Bingham and the German-born Karl Ferdinand Wimar.

Hastings was educated at St Louis University and then worked as a commercial sign painter and decorator. In the late 1850s, he travelled to Düsseldorf, Germany to study art under the English artist Joseph Lindo and to Paris where he copied paintings in the Louvre. Hastings, a vigorous individual, died aged 84 after a productive career: he once boasted that he had painted one picture per month for 50 years. Among them were at least 13 depictions of St Louis volunteer fire companies in action or on parade.

Targee (1808–49) was Captain of the Missouri Valley Fire Company No. 5 on the night of 17 May 1849 when a major conflagration occurred in the river port. A steamboat caught fire and then set alight 22 other boats and the waterfront with its stores and warehouses full of inflammable goods. The fire threatened to consume the whole of the city. Unable to halt it by conventional

100 Matthew Hastings, *Captain Thomas Targee*, (1902). Painting, oil on canvas, 123 x 92 cms. St Louis, MO: Missouri Historical Society. Reproduced courtesy of the MHO

means, the fire chiefs decided to create firebreaks by the ancient method of using gunpowder to blow up buildings. After six had been destroyed, Targee introduced a keg of gunpowder into Nathaniel Phillips' music store, which was already on fire, and was killed by the ensuring explosion. He was the first fatality of the St Louis volunteer firefighters. His grave is located in Bellefontaine Cemetery, St Louis, and the street where he lived was renamed after him.

Hastings was 15 when Targee was killed. In later years, he distinctly recalled the 1849 fire and Captain Targee's fate. Hastings himself served as a volunteer firefighter for the Missouri Company and later became an honorary member of the Veteran Volunteer Firemen's Historical Society of St Louis, which commissioned the painting in the late 1890s. Hastings' somewhat belated tribute to a local hero was unveiled at the annual meeting of the Missouri Historical Society in 1902.

In Hastings' rendition, Targee stands bareheaded in the middle of a wide street and adopts a commanding frontal pose. One wonders if Hastings had a likeness of Targee because, strangely, his face strongly resembles that of the painter Bingham. The imposing figure and dark clothing forms a powerful silhouette. Targee holds a top hat in his right hand, grasps a speaking trumpet in his left and points towards a row of burning buildings. Behind him, is an animated scene in which crowds of firemen operate manual pumps, move hosepipes, carry ladders and direct jets of water. There is a dislocation between figure and background, which implies that Targee – who is a giant compared to the other figures – is already separated from his fellows by the sacrifice he is about to make.

101 Gail Sherman Corbett, *Fireman*, (1905). Memorial/public sculpture, bronze, life-size. Syracuse, New York: Fayette Firefighters' Memorial Park. Photo A. Pierce Bounds

## 101 1905

## Gail Sherman Corbett, Fireman

This handsome, young, bronze fireman seated in a relaxed pose holding his fire hat in his lap was sculpted in a social realist manner by Gail Sherman (later Mrs Corbett, 1871–1952) in the early 1900s. The Henry-Bonnard Bronze Company of New York City cast the figure. Apparently, Sherman was a controversial choice for the commission because of her lack of experience (and no doubt because of her gender); however, the fact that she was an artist of local origin counted in her favour. Sherman was born in Syracuse, New York, studied at Syracuse University and then trained at the Art Students' League in New York and at the École des Beaux-Arts in Paris. For a time, she was also a pupil of the prolific and commercially successful Irish-American master sculptor Augustus Saint-Gaudens (1848–1907). Sherman was clearly a skilled modeller of the human figure and she went on to execute a number of allegorical public sculptures.

Her fireman figure, which symbolises quality and fortitude, is part of a larger memorial commemorating Hamilton Salisbury White (1853–99), a Syracuse inventor and President of the Board of Fire Commissioners, who died from a heart attack on 13 March 1899 after attending a blaze in a South Salina Street store. He had provided Syracuse with its first fire engine and fire alarm boxes, and had maintained a fire company at his own expense. A bronze bust of White rests on top of a pink granite pedestal flanked on one side by the young fireman, and on the other by a bronze mother who holds a boy in her lap. Since the boy holds a toy fire engine, it is clear what his future occupation is likely to be. Two curved stone benches on either side of the central pedestal complete the monument, which had been commissioned in 1900 and its $12,000 cost raised by donations from local firefighters and citizens who included many children.

Sherman selected the prominent American architect, teacher and writer Harvey Wiley Corbett (1873–1954) to design of the monument's base. They had met as art students in Paris and fallen in love while both were living in New York. The couple were married the day after the monument's unveiling on 27 June 1905.

As mentioned earlier, also located in Fayette Park is another firefighter memorial dedicated to Philip Eckel (1827–86), a man of German origin who was Chief Engineer of the Syracuse Fire Department. Carved from Westerly granite, it dates from 1900. Eckel died on 1 June 1886 when he was thrown from a ladder truck en route to a fire. The figure wears a fireman's uniform and cap, holds a speaking trumpet under his left arm and stands on a pedestal made from Barre granite, which has a bronze plaque and four fire hydrants. The sculptor responsible for the statue is not known but the contractors were Carrick Brothers and Emslie, McLeod & Company. Originally, the memorial was sited on the City's north side but was moved in 1959 and again in 1979 to its present location.

A third firefighters' monument constructed from granite – *The Collins Block Memorial* – was added to the Park in 1939. It commemorates nine Syracuse firefighters who died when the downtown Collins Block collapsed on them after they had extinguished a fire. Like the White monument, it has a central vertical block flanked by curved stone benches but this time there is a relief carving on the front face of the central block depicting a firefighter with a hosepipe. William Cowie was the sculptor and the fabricators were Sanborn Granite Company and Cook, Watkins, Patch.

## 102 — 1909

### Caspar Buberl/Fiske Iron Works, Fireman's Drinking Fountain

## 103 — 1908

### Giuseppe Moretti, Jacob Livingston Memorial Statue

In terms of its iconography – a fireman carrying a child and holding a lantern – the Slatington statue is conventional. What is unusual about it is that it has been painted in several rich colours and that it crowns a drinking fountain for dogs, horses and humans. It is presently located on Main Street just outside the Public Library. It was purchased from J.W. Fiske Iron Works, New York for $700 in 1909 and dedicated on 23 April 1910 by Slatington's Hose Company No. 1, which had been founded in 1885. An old Fiske catalogue illustrates the statue, No. 500 D, and credits the artist 'Buberle'. This must surely refer to Caspar Buberl (1834–99) a Bohemia-born sculptor who studied in Vienna and came to New York in 1854 where he established a studio. At night, the fireman's lantern is illuminated via an electric light bulb. Today, the fireman figure has a red-fronted helmet and his double-breasted belted coat and trousers are coloured blue. The infant cradled in his left arm wears a long white nightdress. This statue symbolises 'service, vigilance and humanity' and has been described as Slatington's 'V.I.P'.

102 Caspar Buberl/Fiske Iron Works, *Fireman's Drinking Fountain*, (1909). Memorial/public sculpture, zinc and paint, c. 2 m high. Slatington, Pennsylvania. Photos courtesy of David D. Altrichter

During the twentieth century, there seems to have been a minor vogue in the United States for polychrome statues placed out of doors on drinking fountains because other painted figures of firemen with infants and lanterns were erected in Owego and Middletown (New York), and Ottawa (Ohio). Owego's *Fireman's Fountain* (aka *Baker Memorial Fountain*) was dedicated on 26 August 1914 to the memory of former Chief Engineer George Hobart Baker who was killed in an automobile accident in 1913. His father, also a fire chief, paid $2000 for the Fountain and donated it to the Village of Owego. It is located in Courthouse Square. According to one American art inventory, the sculptor responsible for the zinc statue was Karl Bitter (1867–1915), an Austrian artist trained at the Vienna Academy of Fine Arts, who came to the United States in 1889. J.W. Fiske Iron Works supplied the Fountain's iron basin. Following a restoration, members of the Baker family attended a re-dedication ceremony in July 2000. Today, the fireman has a red front to his helmet and wears a light blue coat, while the Fountain on which he stands is painted green.

Ottawa's *Volunteer Firemen Memorial* located near Putnam County Courthouse was dedicated on 25 November 1915 to honour the town's volunteer firemen and in particular John Krieger and Joseph Hermiller, Jr., two local firemen who died on duty in 1911. The fireman-and-infant statue, sculpted by Buberl and made from cast zinc painted gold, was placed on a Victorian-style rec-

103 Giuseppe Moretti (designer), *Jacob Livingston Memorial Statue*, (1908). Public sculpture/drinking fountain, zinc with iron base, J.T. Mott Iron Works & A.T. Shupe Marble & Granite Dealers. Carlisle, Pennsylvania. Photo courtesy of Richard Tritt, photo curator of the Cumberland County Historical Society

tangular plinth set in a low fountain basin. This statue is identical to those of Owego, Middletown and Slatington, and was supplied by the same firm, Fiske Iron Works; consequently, Fiske must have had a mould capable of generating many examples of the same sculpture.

Yet another painted fireman-with-infant fountain can be found in Carlisle, Pennsylvania. It is similar to those already described, but the fireman is clean-shaven and his coat is open. The cast zinc statue, from a design by Giuseppe Moretti, was made by J.T. Mott Iron Works of New York and erected by A.T. Shupe Marble & Granite Dealers after being commissioned by Mr and Mrs Jacob Livingston in memory of their daughter who had died in 1901. (Jacob was a local tobacco retailer.) The fountain, which cost $3000 and was donated to Carlisle Borough, was erected in a public square in front of the local courthouse in December 1908 and dedicated in October 1909. Subsequently, it was moved several times but since 1984, it has been located in front of 102 West Ridge Street Fire Company. During a restoration undertaken in 1999, the statue was given a new head and painted bronze, which had been its original colour. However, as the photograph shows, during the past, it was multi-coloured.

Museum goers familiar with white marble sculptures of ancient Greece and Rome may be surprised to learn they were once painted in bright colours; hence, the practice of colouring statues is a venerable one. During the 1970s, the hyperrealism of coloured sculptures was manifested in the work of such American sculptors as John de Andrea and Duane Hanson. Therefore, in this instance, there is an affinity between a tendency in modern art and the statuary associated with firefighting

## 104 — 1908

### Fred S. Howard, Escape Turn out

## 105 — 1910

### Unknown artist, New York Fire Department

Picture postcards as a popular medium for sending personal messages and greetings date from the late 1890s when postal authorities in Britain and the United States first permitted their use. They also depended on printing technologies that enabled the mass production of images in black and white, sepia or full colour. Of course, many cards feature photographs while others reproduce drawings and watercolours generated by artists and cartoonists. Among the millions of postcards produced since 1900 that have survived and been obsessively collected, many thousands depict fires, fire engines, stations and brigades.

In the first image, a team of five brass-helmeted firemen of the London Fire Brigade respond to an alarm by ‘turning out’ their wheeled ‘escape’ [ladder], which is mounted on a cart pulled by two greys, from the

104 Fred S. Howard, *Escape Turn out*, (1908?). Colour picture postcard, 9 x 14 cms. London: Misch & Co., The Fire Brigade Series No. 434. Author’s collection

105 Unknown artist, *New York Fire Department* – 1244, (1910?). Colour picture postcard, 9 x 14 cms. New York: A.C. Bosselman & Co., postmarked 1910. Author's collection

entrance of a fire station. The illustrator, Fred S. Howard, has reverted to the side view first used in the 1830s. The church in the background resembles James Gibbs' Baroque-style St Mary-Le-Strand (1717) and therefore the fire station on the left is probably located near it. Misch & Co, the publishers, began selling postcards in Britain from 1904 and since this particular card is postmarked 1908, it must have been produced between those two dates. The card is one of a series about the Fire Brigade, so it probably appealed to collectors desiring a full set.

While the fireman-driver holds the reins and a whip, a comrade holds a brake lever. This particular team of firemen specialised in rescuing victims rather than in putting out fires although, of course, once a ladder was in place it could be used by hose companies. On the side of the cart are the letters LCC, standing for the London County Council, the body which took charge of the Metropolitan Fire Brigade in 1889 and altered its name from 'Metropolitan' to 'London' in 1904. (The fire station's façade retains the initials MFB.) As buildings in major cities became taller, nineteenth-century escape ladders were designed with extensions so that they could reach higher. (The red-painted one depicted in the postcard could reach 15 metres.) However, this postcard captures a passing era because soon motors would replace horses and petrol engines would drive escape ladders previously cranked by hand. Wooden ladders would also be replaced by steel ones and mounted on turntables on the backs of trucks so that they could revolve around a 360–degree axis.

The second postcard depicts the arrival at a night-time fire in New York of a steam-powered fire engine drawn by three, wild-eyed, white horses. Its driver must have departed in haste because he is without a cap or helmet. A fire chief dressed in a dark waterproof coat points to where he wants the engine to go. On the left, a steam engine is already pumping water towards the flames in the background and in the far distance, firemen are climbing a ladder. In the middle distance, a water tower engulfed in smoke is delivering a jet of water from on high. This image was painted with more verve that the British one and is somewhat expressionist in character. The artist who created it is not credited but may have been German because, although the postcard was published by a New York company, it is marked 'Made in Germany'.

## 106 — 1909

### William Wallace Denslow, The Fireman

W.W. Denslow (1856–1915), a leading Midwestern American cartoonist, illustrator, book cover, costume and set designer, and writer, was born in Philadelphia but raised in Manhattan. He studied art at the Cooper Union and the National Academy of Design in New York during the 1870s and subsequently obtained magazine and newspaper work in Chicago and San Francisco. Denslow became famous for illustrating *The Wonderful Wizard of Oz* (1900), L. Frank Baum's best-selling chil-

106 William Wallace Denslow, *The Fireman*, (1909). Lithographic illustration, 19 x 13.5 cms. Author's collection

dren's book. As a person, Denslow has been described as 'colourful, brash and cynical ... a large man with a walrus moustache and a foghorn voice'.

This charming colour illustration of a fire chief barking orders to his men while standing precariously on two beams above flames first appeared in the children's magazine *St Nicholas* in April 1909. It was one a series Denslow created about occupations entitled *When I Grow Up*. In September, The Century Company of New York published the series in book form. (This volume is now an expensive collectors' item.) Rather strangely, the chief prefers to shout rather than use the speaking trumpet in his left hand. Behind him, a row of four firemen carry and direct a hosepipe. Their rather small statures must have encouraged young readers to identify with them. The illustration's areas of flat colour enclosed by black outlines reflect the influence of Japanese prints that Denslow admired. On the right-hand side of the image is the artist's famous trademark signature in the shape of a stylised seahorse, while on the reverse is a black-and-white drawing of a fire engine driver and two horses racing along plus the following lines: 'Our coachman's boy – called "Tiny Tim" – is never afraid. When he grows up, he's going to be "Chief" of a Fire Brigade. Says Tim: 'I'd wear my boots to bed, When I'd lie down to rest; And at the striking of the gong I'd be the first one dressed!'

A collection of Denslow's drawings and papers is preserved in the library of the University of Mississippi and a biography written by Douglas G. Greene and Michael Patrick Hearn was published in 1976.

## 107 — 1908

### Unknown illustrator, Saved!

## 108 — 1900–1910

### Unknown illustrator, Fighting the Flames, American Scaling Ladder

Two vividly coloured illustrations of the familiar rescue or saved motif, both of which employ a quasi-expressionist style. The first, which depicts a fireman emerging from a window holding in one arm a doll-like girl, appeared in the 16–page children's book *Story of the Firemen* (aka *Story of the American Firemen*) produced by the McLoughlin Brothers of New York, copyrighted 1908 but may have been published in 1909 or 1911. (In 1985, it was reprinted by B. Shackman & Co., Inc., of New York.)

The publishing house, which specialised in children's books, comic almanacs, games and valentines, was founded in 1828 by John McLoughlin, a Scotsman, who arrived in New York around 1819. The firm was continued by his son and grandsons and became highly successful. Many illustrators were employed and credited by the publishers but, unfortunately, in this instance, none of the images are signed nor is the author of the text credited.

107 Unknown illustrator, *Saved!* (1908?). Chromolithograph book illustration, 24 x 18 cms. New York: McLoughlin Bros., publisher

The publisher of the picture postcard – Raphael Tuck & Sons, founded in 1866 – was art publisher to Queen Victoria and later became one of Britain's largest manufacturers of Christmas cards and postcards. Its 'Oilette' type reproduced paintings, either existing ones or specially commissioned ones. In this instance, the subject matter is American and at least two rescues are in progress involving young women trapped inside a blazing building.

What the two illustrations have in common is that both depict the use of hook or scaling ladders. Historians claim they were a French invention; hence, they were also called 'pompier ladders'. Foreman Chris Hoell of St Louis Hook & Ladder No. 6 patented an American version during the early 1880s, which was first used in New York in 1884. Such ladders were devised in order to reach beyond the tops of conventional ladders. They became necessary as the height of buildings increased.

The ladder had a single vertical beam pierced by horizontal rungs and a curved serrated hook at the top end that was placed over a window ledge and so a fireman could climb up, straddle the ledge, raise the ladder and

108 Unknown illustrator, *Fighting the Flames, American Scaling Ladder*, (1900–10?). Picture postcard, 9 x 14 cms. London: Raphael Tuck & Sons, 'Oilette', 'Fighting the Flames' Series II.

ascend to the window above, thus proceeding up the façade of the building floor by floor. Using a rope or lifeline attached to his belt, the fireman could then lower victims down. Naturally, this rescue method was highly risky and required great skill and strength. The subject of the British postcard may have been prompted by the fact that a team of firemen from Kansas City won a pompier ladder competition at an International Fire Congress held in London in 1893.

## 109 — 1909

### Unknown artist/photographer, A Typical Fireman

This British postcard dating from the early years of the twentieth century reproduces a head-and-shoulders, three-quarter-view portrait of a London fireman. His crested brass helmet features the letters LCC (London County Council) and MFB (Metropolitan Fire Brigade). His fine facial features, pink complexion and blue eyes, handlebar moustache and beard, plus his alert pose, steadfast gaze and polished brass buttons signify a noble, ideal, white, masculine character who is proud of his uniform and occupation. (The man's face resembles that of the young King Edward VII, who as Prince of Wales was a keen amateur fireman.) The card's title '*A Typical Fireman*' claims that he epitomises all of London's firemen. In short, this portrait is a powerful and persuasive example of pictorial rhetoric designed to appeal to a popular audience and to postcard collectors.

In 1851, James Valentine (1814–79), a Scottish engraver and photographer, founded the 'photographic publishing' company Valentine & Sons in Dundee. When James died, his two sons William (1844–1907) and George (1852–90), both of whom acquired photographic skills, continued the business. By 1900, it was using the collotype lithographic printing process to produce picture postcards on a large scale. ('Artotype' is an alternative term for collotype.) Seven years later, the

109 Unknown artist/photographer, *A Typical Fireman*, (1909?). Colour picture postcard, 9 x 14 cms. Dundee: Valentine's 'Artotype' series. Card postmarked 1909. Author's collection

company was employing around 40 artists to retouch negatives. A collection of Valentine photographs – which includes three portraits of firemen – and postcards, mostly topographical, is preserved in St Andrew's University Library.

## 110 — 1910

**Unknown illustrator, A Truck of English Suffragettes Disguised as Firefighters**

## 111 — 1910

**Unknown cartoonist, Flossie, the Firegirl. Displaying her Hose**

1910 was an election year in Britain and a time when women demonstrating for the vote and other rights were being arrested, imprisoned and force-fed. Some militants adopted the habit of setting post-boxes alight and making false fire alarms. In April 1913, some suffragettes burnt down the mansion called 'Levetleigh' in St Leonards, Hastings, which was the home of Arthur du Cros, a Member of Parliament opposed to the 'Votes for Women' campaign. (The house was empty at the time.) The campaign finally succeeded in 1918 in recognition of women's contribution to the war effort during World War I.

A colour illustration published in a supplement to the popular French newspaper *Le Petit Journal* (established by press magnate Moise-Polydore Millaud in 1863) on 3 April 1910, by an unknown artist, depicts English suffragettes taking part in a London street parade dressed in firefighting helmets and uniforms (some of their own devising). Five women ride on a horse-drawn ladder wagon, which is preceded by two 'cadets' skipping gaily along. They are waving flags and carrying umbrellas adorned with the slogan 'Votes for Women'. Exceptionally, one flag announces a meeting at the Albert Hall. The illustration's caption makes it clear the firewomen were in disguise but the implication was that, given equal rights and opportunities, women could perform tasks traditionally associated with men. However, it is ironic that while some women posed as firefighters, others were setting fires!

George du Maurier (1834–96), a French-born British artist, cartoonist and novelist, anticipated this illustration when he created an image of a horse-drawn cart with an all-female fire crew for *Punch* magazine, 12 May 1866. His image, entitled '*The last fast thing*', satirised 'fast' women who tried to equal men.

A possible source for the French magazine's illustration was a press report about the Burton Ladies Fire Brigade (that is, Burton upon Trent, Staffordshire), which used to provide demonstrations at the annual local Whit Monday Fete and Gala around 1910–11. Watched by large crowds, the 18 members of the Brigade used to drill and race around a field. Postcards and a film of their activities were soon released. It seems they never fought any real fires and that their displays were for entertainment only.

The dress of the 'cadets' in the French illustration resembles that of six women led by a Miss Mortimer who visited Paris and Vincennes in 1889 to attend an International Congress of Officers and Sub-Officers of Fire Brigades. Ronald Cox, Shaw's biographer, describes them as 'connected to the theatrical profession' and says they were 'to exhibit and advertise a special life-saving apparatus', namely a canvas chute, which they demonstrated to the cheers of a watching crowd. Male viewers caught glimpses of female legs as they emerged from the chute feet first.

Events such as these surely prompted the production of the saucy postcards that were published in Britain during the early twentieth century. For instance, one shows a 'firegirl' dressed in a blue uniform edged with red and a yellow brass helmet, climbing a ladder while

110 Unknown illustrator, *A Truck of English Suffragettes Disguised as Firefighters*, (1910). Illustration, 40.5 x 29 cms, published in *Le Petit Journal, Illustrated Supplement*, 3 April 1910. Photo courtesy of the Brotherton Library, University of Leeds

111 Unknown cartoonist, *Flossie, the Firegirl. Displaying her Hose*, (c. 1910). British Postcard, 14 x 9 cms. Author's collection

hauling a hosepipe. The hem of her skirt is just below her knees. At the base of the ladder, a trio of male voyeurs ogle her calves, ankles and petticoat. The punning caption reads: 'Flossie, the firegirl. Displaying her hose.' Perhaps to avoid any adverse attention from censors, the names of the artist and the publisher are not credited on the back of the card. (The card reproduced was never posted and so there is no date or message on the back.) The artistic style of the image resembles that of Donald McGill who began his long career in 1904 and was to become famous for his naughty seaside postcards made for the company Bambforth of Holmfirth, Yorkshire. Clearly, the novelty of female firefighters made them subjects for jokes and puns; they also stimulated the erotic imagination of many men.

Documentary photographs in the Hulton-Getty archive dating from March 1916, depict women in firefighting gear drilling and performing exercises with pumps and ladders in streets watched by crowds of bystanders; consequently, it seems some women were recruited as firefighters during World War I to replace firemen who had joined the armed forces. Further images from the same archive reveal that, in the 1920s, private companies such as the dyers and cleaners Achille Serre Ltd had all-female fire brigades equipped with uniforms, brass helmets, axes and portable fire extinguishers. However, it was not until World War II that thousands of women were recruited for the Auxiliary Fire Service.

## 112 — 1910–1911

**Unknown sculptor, Gargoyle on an ex-Firehouse No. 9 at 2 South 14th Street, Kansas City, Kansas**

## 113 — 1920s

**Unknown sculptor, Gargoyle on an ex-firehouse at 1328 Race Street Philadelphia**

As we have seen, two- and three-dimensional sculptures of firemen and scenes of firefighting have appeared on memorials in cemeteries. Another common site in which they can be found are the façades of fire stations. Some sculptures adorning American ones have taken the form of firemen-gargoyles. For example, the principal corners of Fire Station No. 9 of Kansas City, Kansas, designed by local architect William E. Harris in 1910–11, were trimmed with terracotta heads of dwarfish firemen wearing helmets with the number nine on them.

This imposing, three-storey, two-bay building made from brick and masonry ceased to be a firehouse in 1967 and is now the Prescott Neighborhood Center. In 1983, it was declared a Historic Landmark. Another ex-firehouse in Philadelphia at 1328 Race Street is even more lavishly adorned with firemen-gargoyles because it has six situated below Gothic-style arches. (Strictly speaking, these sculptures should be called 'grotesques' rather than 'gargoyles' because the latter term refers to

112 Unknown sculptor, *Gargoyle on an ex-Firehouse No. 9 at 2 South 14th Street, Kansas City, Kansas*, (1910–11). Photo courtesy of Robert Rocha

113 Unknown sculptor, *Gargoyle on ex-firehouse at 1328 Race Street, Philadelphia*, (1920s?). Photo courtesy of Northstar Gallery

carvings on waterspouts.) The building was in use as a firehouse and acted as the headquarters of the Philadelphia Fire Department from 1926 to 1976. Engine Company No. 17, Truck Company No. 23, Pipeline No. 5, and Heavy Rescue No. 1 occupied it at a time when horse-drawn fire engines were being replaced by motorised ones. The brick building with its Gothic-revival features must have been constructed before 1926 but, strangely, the architect is at present unknown. The artist or stonemason who carved the sculptures is also unknown and the reason for their creation, although, clearly, they paid homage to the iconography of Europe's medieval cathedrals.

Compressed between Gothic arches and brick columns, the stone helmeted figures squat like miners in a tunnel. Their caricatured faces have expressions ranging from pleasure to pain, from determination to fear. Large, powerful hands grip axes, hoses and someone being rescued. The primitivistic carvings are crudely finished – claw-tooth chisel marks are still visible – but this is refreshing because it endows them with an energy and liveliness rare in firefighting sculpture.

## 114a/b 1912–13

### Harold van Buren Magonigle and Attilio Piccirilli, Firemen's Memorial

Overlooking the Hudson River at West 100th Street and Riverside Drive, New York, is the *Firemen's Memorial*, which was erected in 1912 and unveiled in 1913. This massive structure cost $90,500 and was a gift to New York from the Firemen's Memorial Fund Committee. It was prompted by the heroism of Charles Washington Kruger (aka 'Big-hearted Charlie'), a veteran fire chief, who died aged 36 on 14 February 1908 while fighting a sub-cellar fire on Canal Street. At his funeral service, Bishop Potter proposed a monument to honour the City's fallen firemen. The architect Harold van Buren Magonigle (1867–1935) designed an oblong, pink marble sarcophagus or cenotaph and the artist Attilio Piccirilli (1886–1945) provided a bas-relief. The Memorial also has an approach of steps, a platform with a balustrade and a drinking fountain basin.

Magonigle, a successful architect, sculptor and writer, was born in New Jersey and earned high respect for his neo-classical style buildings and monuments. Flanking the cenotaph are two huge stone carvings, credited to Magonigle, of seated figures personifying 'Sacrifice' and 'Duty'. Sacrifice is a pietà-type sculpture: a partially draped female supports the semi-naked body of a dead fireman across her lap. Duty, a widow frozen with grief, sits next to a fire hydrant and consoles her young son with one arm while holding her late husband's helmet with the other.

114a 114b Harold van Buren Magonigle and Attilio Piccirilli, *Firemen's Memorial*, (1912–13). Monument/public sculpture, Knoxville marble, Milford granite and bronze; sarcophagus: 6 x 7.5 x 2.4 m. New York: Riverside Drive. Photos: Lee Sandstead

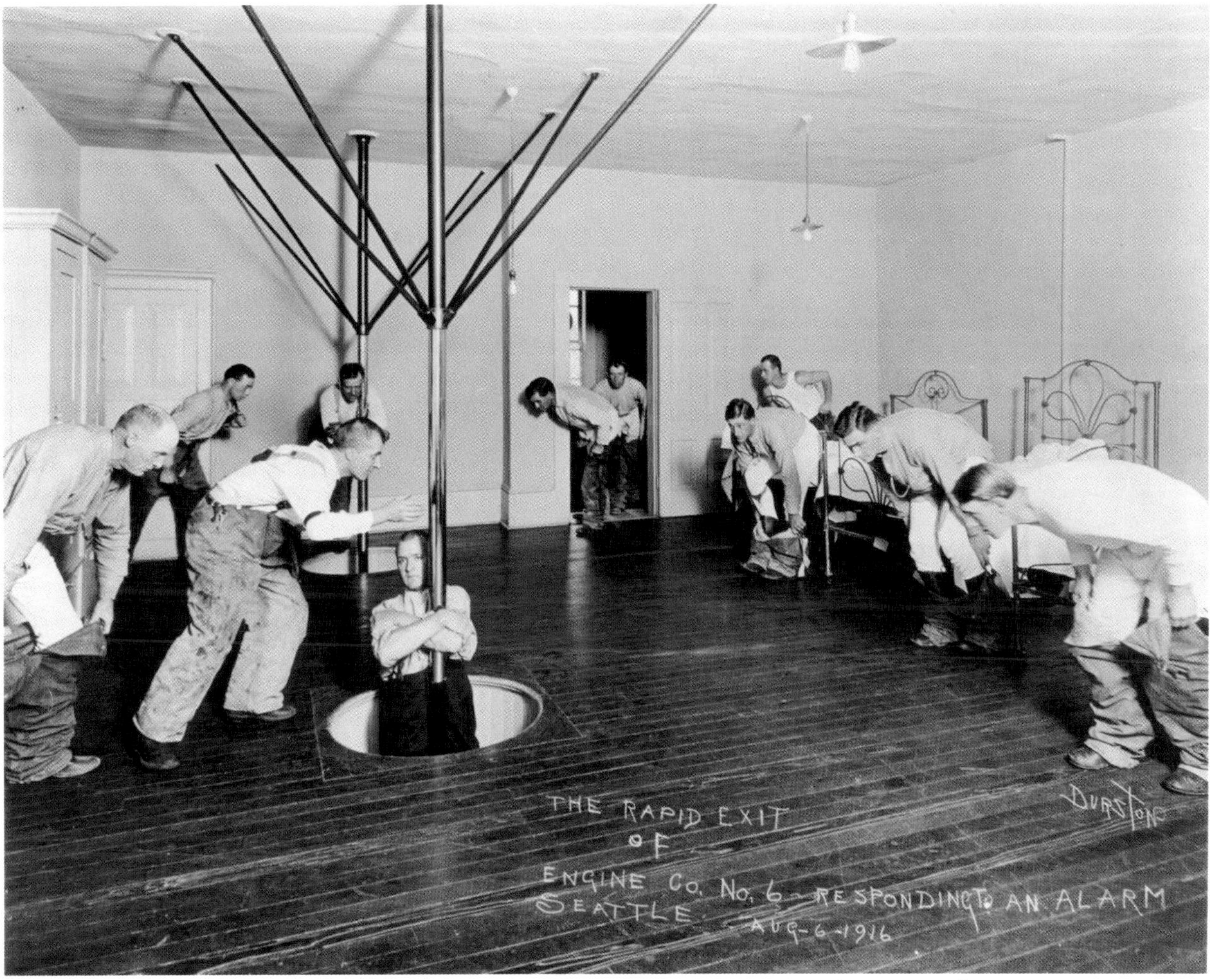


115 Durston, *The rapid exit of Engine Co. No. 6 – responding to an alarm, Seattle,* (6 August 1916). Photo courtesy of the Museum of History and Industry, Seattle

According to art historian Lee Sandstead, Piccirilli was a member of a family of Italian-American marble cutters and sculptors, which first arrived in the United States from Massa-Carrara, Italy in 1888.

For over 40 years, the Piccirilli Studio (founded 1893) in the Bronx, which specialised in carving large stone statues based on small models, was the biggest in America. In this instance, therefore, there was a direct line of tradition to the skills and figurative sculpture of the Italian Renaissance.

Adorning the west face of the cenotaph is a large bronze relief depicting three galloping horses pulling a fire engine. (Originally, this relief had been made from marble but, due to weathering, it was replaced with a bronze version during a 1934 restoration. A further restoration was undertaken by Jan Hird Pokorny Associations Inc., New York, from 1987 to 1992.) Thus, the Memorial combines a scene of early twentieth century firefighting with Renaissance-style, female personifications. On the east face is a relief depicting the mythical bird Phoenix plus a dedication, composed by Royal Cortissoz, to the 'soldiers' who fight 'a war that never ends'. In addition, the cenotaph is ornamented with carvings of flames and scrolls.

The Memorial is dedicated to all the fallen firefighters of New York (now numbering over 1000) and remembrance services take place there annually, which the City's Mayor and Fire Commissioner normally attend. American art historian Donald Martin Reynolds regards the Memorial as 'a superlative union of architecture, sculpture and symbolic ornament' and as New York's 'quintessential commemorative monument'.

In 2002, the American giftware and collectibles manufacturers Department 56, of Eden Prairie, Minnesota added a miniature replica of the Memorial to its American Pride Collection. It is made from resin composition, stands on a marble base and is accompanied by a patriotic pin and a scroll. Obviously, this household ornament, which measures 15 x 11 cms, cannot rival the scale and public presence of the original.

116 *"Firemen, Save my Child"*, (1927). Cover of four-page publicity leaflet or movie herald, 11.5 x 15 cms. Hollywood: Paramount Pictures. Author's collection

| | |
|---|---|
| **115** | **1916** |
| **Durston, The rapid exit of Engine Co. No. 6 – responding to an alarm, Seattle** | |
| **116** | **1927** |
| **"Firemen, Save my Child"** | |
| **117** | **1958** |
| **Gil Elvgren, Fire Belle** | |

During the nineteenth century, hundreds of fire stations were built in Europe and North America to accommodate men and appliances. (In 1982, a substantial history and survey of the architecture of American firehouses, written by Rebecca Zurier, was published.) Once professional paid firemen were spending days and nights on duty or 'watch' inside fire stations, design logic dictated that horses and appliances were located on the ground floor and the men's living and sleeping quarters on the first floor. Some stations had three storeys, with oats and hay stored on the first or third floors. An oval-shaped, nineteenth-century magazine illustration, published in *Harper's Weekly*, (4 March 1876), showing the bustle inside the firehouse of the New York Insurance Patrol following an alarm, depicts firemen pulling on their clothes while hurrying down a long staircase in order to reach horses waiting to be hitched to appliances. It seems horses sometimes tried to climb the stairs and so spiral ones were introduced but these slowed the firemen down. (In some cases, chutes instead of staircases were used.) An additional corner illustration shows the men's living quarters with a row of beds.

Clearly, any device that enabled firemen to reach the ground floor more quickly was welcome. Enter the sliding brass pole that was introduced in the United States in the 1880s. Britain soon followed suit. In fact, according to firefighting historians, the first pole was a 7.5 cms diameter, polished one, made from Georgia pine, introduced by Captain David B. Kenyon of Chicago's Engine Company No. 21 in 1878. (This Company was all African-American.) Although Kenyon's pole became the butt of many jokes, No. 21 began to reach fires before other companies and so its value was soon recognised. The first brass pole dated from 1880 and was devised by Captain Charles Allen of the Worcester (MA) Engine Company No. 1. The sight of a fireman – often half dressed – descending the pole, viewed from the side or from above or below, soon became a recurrent pictorial motif. It also proved to have comic and erotic potential.

Even documentary photographs of pole descents could be amusing, as the one showing Seattle firemen hastily pulling on their pants during an alarm in August 1916 demonstrates. The comic potential was fully exploited in the 1927 Paramount silent film *"Firemen, Save my Child"*, directed by A. Edward Sutherland and starring Wallace Berry and Raymond Hatton as Elmer and Sam, two feckless firemen. (Incidentally, no less than six films were made during the period 1918 to 1954 with the title '*Firemen (or Fireman), Save my Child*'.) The funniest moment in the film occurs when a fire captain slides down a pole to answer an alarm only to discover that its lower end has been 'borrowed' by Sam.

More jokes about a fire pole occur in the 1939 British comedy film *Where's that Fire?* directed by Marcel Varnel and starring the famous radio, music hall and screen comedian Will Hay (William Thomson Hay, 1888–1949) as Captain Benjamin Viking, an incompetent officer at the antiquated Bishop Wallop fire station. One reviewer describes scenes of 'pandemonium' and 'escalating public frenzy' as Viking and his mates attempt to erect a new pole and in so doing cause a traffic jam, wreck a china shop, terrorise a bedridden invalid and wreck his house.

The erotic potential of the motif manifested itself in fantasies of scantily dressed young women clinging to shiny poles. (Lascivious pole dancing is an attraction of

117 Gil Elvgren, *Fire Belle*, (1958). Painting/illustration. (c) Brown & Bigelow, Inc., Saint Paul, Minnesota

many contemporary strip clubs.) One such image was created in 1958 by the highly successful and prolific American pin-up and glamour artist and photographer Gil Elvgren (1914–80). His 'creamy, smooth as silk' figure paintings were reproduced as illustrations in countless books, calendars, magazines, playing cards and advertisements. Entitled *Fire Belle*, it depicts an open-mouthed, curvaceous blonde descending a pole while wearing a black lace bustier, sheer white dressing gown, and bright red fire hat and boots.

*Bridget Jones's Diary*, a romantic comedy film directed by Sharon Maguire and starring Renée Zellweger as a hapless singleton, was a box office hit in 2001. In one scene, Jones reports for television from a fire station in Lewisham, London. She then simulates a firefighter by sliding down a pole. Jones is filmed from below in such a way that her generous bottom fills the frame. Of course, since women firefighters now exist, females do literally descend poles – witness the photo-portrait of Eltham Station Commander Danielle Cotton taken in 2002 (see illus 214). Despite the fact that the masculine sexual fantasy is now both outmoded and inappropriate, the presence of women in fire stations prompts romantic plots in television drama series such as *London's Burning*.

## 118 — 1916

## Charlie Chaplin, The Fireman

118 Charlie Chaplin, *The Fireman*, (1916). Feature film, silent, black-and-white, 20–min two-reeler. Hollywood: Mutual Film Corporation. Photo: detail from a poster for the film. Author's collection

At first sight, firefighting might seem too serious a subject for humour but a number of comedy films have been made about it. (Although, obviously, there is much potential hilarity in things like the water jets of hosepipes.) The English-born music hall and screen star Charlie Chaplin (1889–1977) made one of the earliest in 1916 entitled *The Fireman* (aka *The Fiery Circle* and *A Gallant Fireman*). He starred in it, produced and directed it and co-wrote the script. At this period of his career he had complete artistic control and was generating comedy films for the Mutual Film Company of Hollywood at the rate of almost one a month.

Charlie plays the part of a clumsy and lackadaisical apprentice fireman at a ramshackle small-town firehouse. When he sleeps through an alarm, his Chief – played by the huge and fearsome-looking actor Eric Campbell – disciplines him by kicking his backside. There follows several scenes of knockabout farce involving Charlie, the Chief, other firemen and a cook plus sliding poles, a fire axe, water and food being thrown about. An affluent property owner visits the firehouse with his attractive daughter and persuades the Chief to let his villa burn down so that he can collect the insurance money, and promises in return that the Chief can marry his daughter. Charlie also falls for the girl played by his regular co-star Edna Purviance.

A man discovers his home is on fire and sets off a street alarm but there is no response from the firehouse because Charlie is busy playing a board game. (To modern eyes, his behaviour is criminal rather than amusing.) Finally, the fire company turns out but, predictably, when it reaches the burning house Charlie douses his comrades with water instead of the flames. Meanwhile the affluent property owner has set light to his villa but then discovers his daughter is trapped in an upstairs bedroom. This time Charlie rushes to the scene driving a steam pumper pulled by two horses but by the time he arrives most of the wagon has been destroyed. Nothing daunted, Charlie scales the front of the villa (there are several convenient footholds) and reaches the damsel in distress. With her unconscious body on his back, Charlie climbs down. Viewers can discern that, during the descent, the 'girl' was in fact a lightweight dummy.

As in so many of his films, Charlie played a bullied underdog who triumphs in the end. This film has been described as 'wall-to-wall slapstick' and although it has some memorable comic moments, it is not one of Chaplin's classics.

119 Wright & Sons, *Firemen's Memorial*, (1924). Memorial/public sculpture, granite, 8 m high. Birkenshaw, West Yorkshire Fire Service headquarters. Photo © 2003 Andrew Hanson and the West Yorkshire Fire Service

Other comedians who have played firefighters on screen include Fatty Arbuckle, Will Hay, Benny Hill, Buster Keaton, Steve Martin, Charles Murray, The Three Stooges and Ed Wynn. One of the finest film comedies about firefighters is Milos Forman's *The Fireman's Ball*, a Czech film made in 1967.

## 119 — 1924

### Wright & Sons, Firemen's Memorial

In 1924, a memorial costing £1200 and paid for by public subscription was erected in Scholemoor Cemetery, Bradford, Yorkshire over the communal grave of six fallen firemen from Odsal Fire Station. They had been killed eight years before on 21 August 1916 following a huge explosion at the Low Moor Munitions Company. It is thought they were crushed when a wall collapsed on their fire engine. In addition, twelve of their colleagues had been injured. Altogether there were 99 casualties, 39 of them fatal. Even though crowds attended a civic funeral for the deceased firemen, news of this tragic accident was restricted because it occurred in the middle of World War I.

The Memorial took the form of single figure of a dignified and stalwart fireman in full uniform holding a hosepipe standing on a rectangular plinth, which was surrounded by a stone balustrade closed by an iron gate. A dedication and list of the names of the deceased appeared on a panel attached to the plinth. A fireman called Tom Farrar modelled for the statue, which was carved from Scottish granite by Wright & Sons, Monumental Masons of Bradford (established in 1875 and still in business). In fact, two statues had to be made because the first was damaged while being transported. It seems the company was almost bankrupted by the additional costs.

In 1999, an off-duty firefighter noticed that thieves and vandals had damaged the Memorial: much of the balustrade had gone and the memorial was disfigured by graffiti. A decision was taken by the West Yorkshire Fire Service to restore the Memorial and to move it to a more protected site near the Service's headquarters in Birkenshaw, Bradford. A sum of £25,000 was raised to pay for the restoration and relocation. A re-dedication ceremony took place in August 2003.

In conception and style, Bradford's *Firemen's Memorial* (aka *Low Moor Memorial*) closely resembles those that were erected in American cemeteries during the second half of the nineteenth century. It is surprising the United States has so many, while Britain has so few.

## 120 — 1927–28

### Avard Tennyson Fairbanks, David Campbell Monument

David Campbell (1893–1911), a Portland Fire Department chief, was killed on 26 June 1911 during a fire at the Union Oil Company plant. A building collapsed on him while he was checking that none of his men were still inside. In 1918, the *Portland Telegram* newspaper encouraged the public to contribute to a fund for the construction of a memorial to Campbell and other fallen firefighters. Campbell's friends eventually raised $35,000 to pay for the Monument, which took the form of a limestone drinking fountain with two lanterns and a bronze plaque. Paul Philippe Cret of Philadelphia designed it while local architect Ernest F. Tucker acted as Portland's representative. It was completed in 1927 and a dedication ceremony took place on 26 June 1928. The small park around the Monument was added in 1963.

Fairbanks (1897–1987) was the sculptor who contributed the bas-relief plaque set into the limestone surface above two fountain basins, which has a fine half-length portrait of Campbell. Like the figures in ancient Egyptian murals, the Fire Chief's body is represented from the front, while his powerful head is shown

120 Avard Tennyson Fairbanks, *David Campbell Monument*, (1927–28). Memorial/public sculpture/fountain, bas-relief plaque, bronze, set in Caen limestone. Oregon: Portland Firefighters Park, West Burnside between SW 18th and 19th Avenues. Photo: Steven Voss, Portland

in a highly defined profile. His overcoat is open to reveal a double-breasted, buttoned uniform jacket and he holds a fire hat in his right hand. This is a striking and memorable portrait of a male American hero as well as being an impressively skilful relief sculpture.

Fairbanks, the son of mural painter, was born in Provo, Utah into a Mormon family and soon manifested a talent for modelling animals. After working at the Metropolitan Museum of Art in New York and at the Bronx Zoo, he studied at the Art Students' League. In 1913, he continued his studies at various academies in Paris. Later, he attended and taught in several American universities. In fact, it was while he was teaching at the University of Oregon that Fairbanks executed the Campbell portrait. Fairbanks became a highly prolific and commercially successful sculptor who undertook numerous public and private commissions for portraits, historical equestrian statues, temple monuments, war memorials, medals, and allegorical and religious sculptures. The latter were created for the Church of the Latter-Day Saints and Fairbanks has been described as defining 'Mormon aesthetics for the greater part of a century'. While his portrait of Campbell is not really an example of 'Mormon art', it is infused with spiritual and humane qualities.

## 121a — 1927

### Unknown sculptor, Firemen's Memorial

## 121b — 1927

### detail

## 122 — 1920s

### Hooper's Studio, Portrait of Fireman G.P. Chamberlain

During the evening of 26 April 1924, a fire broke out on the British steamer *City of Singapore* that was moored for unloading at Dock No. 2 (already known as 'Tragedy Dock') in Port Adelaide, South Australia. The fire was potentially very dangerous because on board the cargo ship were large quantities of oil and petrol. Men from the South Australia Metropolitan Fire Brigade rushed to the scene on their new Dennis fire engines and boarded the ship with their hoses. Within hours they were convinced they had extinguished the blaze. However, at 10.50 pm there was a huge gas explosion that killed three firemen and injured ten others. The dead were Albert Greenman aged 31, George James Anderson aged 23 and James Hickey aged 22. Greenman, who was an ex-soldier who had served on the Western Front during World War I and who drove a fire engine, died from a head wound while the two younger men were at first described as 'missing' because their bodies could not be

121a Unknown sculptor, *Firemen's Memorial*, (1927). Sculpture, life-size, marble. Woodville, Australia: Cheltenham Cemetery. Photo: Joanna Barr

121b Detail. Photo: Joanna Barr

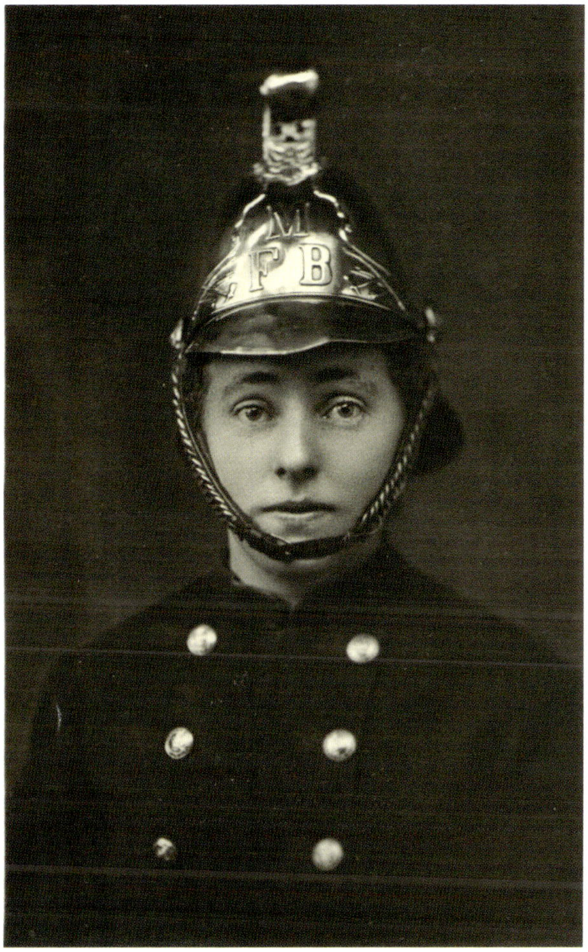

122 Hooper's Studio, *Portrait of Fireman G.P.Chamberlain*, (1920s) Photo/postcard, 13 x 8 cms. Author's collection

found. This was because they had been blown to pieces and only small fragments were later discovered.

Following an elaborate funeral in which fire engines were festooned with wreaths, Greenman was buried in Adelaide's West Terrace Cemetery. Three years later, in March 1927, a memorial to Anderson and Hickey, commissioned by the citizens of Port Adelaide, was unveiled by Mr Frank Condon MP in Cheltenham Cemetery, Port Road, Woodville. It consists of an oblong stone plinth on which is a fine life-size statue of a fireman attired in uniform and helmet who stands proudly to attention and who gazes straight ahead. Italian stonemasons carved the figure from Carrara marble using photographs supplied by the monumental stonemason firm S.D. Tillett Memorials Pty. Ltd. A head-and-shoulders portrait postcard taken by Hooper's Studio, 128 Rundle Street, Adelaide, exists depicting Fireman G.P.Chamberlain. It seems probable that he served as the model for the statue because his face strongly resembles that of the marble figure.

In 1986, Port of Adelaide firefighters were shocked when vandals attacked the *Firemen's Memorial*. In recent years, Joanna Barr of Artlab Australia has inspected the statue with a view to cleaning and restoring it. Although now worn by the elements, this monument continues to commemorate a major disaster in Port Adelaide and the ultimate sacrifice made by two of its young firemen.

## 123 — 1929

**Chicago Daily News photographer, Boy looking toward the camera sitting in a miniature wooden fire truck in front of a fire station**

## 124a/b — 1957

**Hassenfeld Bros Inc., Fearless Fireman, Thrilling Game and Toy**

## 125 — 1907

**Bernard Wall, Little Breeches says, 'I want to be a fireman when I grow up'**

Toys related to firefighting date back to the second half of the nineteenth century when play sets consisting of small-scale models of horse-drawn fire engines, fireman figures and firehouses made from painted wood, cast iron, lead, tin or lithographed paper were manufactured by such companies as Bing in Germany and Pratt & Letchworth and George W. Brown in the United States. Of course, the toys imitated real objects with varying degrees of accuracy and verisimilitude. A popular wind-up toy made by the American company Marx consisted of a fireman who climbs up and down a ladder. One horse-drawn ladder wagon issued in 1903 came with a booklet entitled *How to Become a Fireman*. Picture puzzles, games and dolls were also soon available. In 1935, the Effanbee Doll Corporation added a 'Skippy' fireman composition doll (35 cms high) to its range of patriotic dolls. In 1957, Hasbro (a company founded by two Hassenfeld brothers in Providence in 1923) issued a 'Fearless Fireman, Thrilling Game and Toy' in a cardboard box with an illustrated cover, which included plastic fire trucks, firemen and civilian figures, and a diecut cardboard house on fire. Further examples of such products can be found in Charles V. Hansen's book *The History of American Firefighting Toys*... (c. 1990).

Today, a wide range toys – many made from plastic – aimed at both toddlers and juveniles are for sale, which employ firefighting and rescue as themes. They include so-called 'action figures' equipped with various accessories, model fire boats, Ride 'n' Rescue fire trucks, and water tankers with pumps and sprays. Some British toys are spin-offs from successful television series such as *Fireman Sam* and *London's Burning*. While most such toys

123 Chicago Daily News photographer, *Boy looking toward the camera sitting in a miniature wooden fire truck in front of a fire station*, (1929). Photo from glass negative DN–0089091, courtesy of Prints & Photographs Department of the Chicago Historical Society, Chicago, Illinois.

have traditionally been aimed at boys, girls are also being catered for. The American company Fisher-Price (Mattel, Inc.), for instance, sells a 'Rescue Hero' character called 'Wendy Waters' aimed at children aged between three and eight. This figure is a member of a 'Hydro Team' and so can be filled with water that is then sprayed.

Costumes consisting of helmets, uniforms, axes, air tanks, etc., such as the red plastic outfit 'Let's Pretend Career Costume: Firefighter' made by Learning Curve International, Inc., of Chicago, aimed at three-to-six-year-olds, enable children to perform the role of firefighter. This serves as a rehearsal for the future and as a means of imprinting a positive image of the occupation on young minds. Wearing such a costume – as in the Chicago photograph of a boy at the wheel of a pedal-powered fire truck equipped with a small bell, manufactured by Gendron Inc., during the 1920s – enables an active identification with the role of fireman/driver.

124a 124b Hassenfeld Bros Inc., *Fearless Fireman, Thrilling Game and Toy*, (1957). Digital photos © www.ToyTent.com

Boys have long been attracted to fires and fire departments, and some lads used to assist the volunteer companies of the nineteenth century as runners and lantern carriers. (Naturally, the sons of firemen were especially likely to follow in their fathers' footsteps and often children dressed as firefighters have functioned as fire company mascots.) In the background of the photograph are other boys wearing helmets and watching an event. This was because the Chicago Fire Department had a youth wing – the Chicago Junior Fire Department – for which it organised all kinds of educational demonstrations, visits to firehouses and entertainments, which no doubt included safety messages. As Anne Nelson has remarked, 'Many firefighters seem to have a special affinity for children, which may be partly protective and partly a matter of shared values. "Some kids want to grow up to be firefighters," goes the saying, "but you can't do both."' A detailed examination of the relationship between children and firefighters can be found in David Andrew Lewis' illustrated thesis *Didja' Ever Want To Be a FIREMAN?* (2002).

Junior departments were established across the United States during the twentieth century and some continue to flourish. There are departments for 'Fire Tots' and 'Fire Cadets'. Those admitting teenagers over 14, actually train them so that they can help in the daily work of the fire departments. Junior departments, of course, are a prime source of future recruits.

A number of images exist of children that can be described as 'aspirational' and 'propagandistic'. For instance, the 'Little Breeches' American postcard by artist Bernard Wall that depicts a boy wearing a fireman's helmet using a hosepipe to extinguish a small fire in a field. The caption indicates his desire to become a fireman when he grows up. In the United States, parents can even buy a bib for their babies with this ambition printed on it!

125 Bernard Wall, *Little Breeches says, 'I want to be a fireman when I grow up'*, (1907?). Picture postcard, 9 x 14 cms. 'Little Breeches' series no. 74. New York: The Ullman Manufacturing Co. Postmarked 1915. Author's collection.

## 126 — 1929

### J.B. Taylor, Front cover of The Scout

Many boys like to play with matches and to light fires. As scouts at campsites in the countryside, they can do so legitimately. Sir Robert Baden-Powell (1857–1941, aka 'BP'), a British army officer, founded the Boy Scouts, a voluntary organisation devoted to the education and well being of boys, in 1908 and in the same year, the first issue of *The Scout: The Finest Adventure Weekly* was published. It continued until 1966. On the cover of the 17 August 1929 (Vol. 25, No. 1112) issue is a vivid, two-colour illustration by J.B. Taylor depicting a clean-jawed young British fireman armed with an axe and wearing a gleaming helmet entering a burning building to rescue an unconscious victim who, exceptionally for such imagery, is male. The cover depicts a scene from a fictional fire-and-rescue yarn 'Adventure after all!' by Alan Ross Warwick. The story is one of a series entitled 'Heroes of every day'. Above the cover image is the statement 'Thrills of the fire-fighters!' which turns out to be the title of a factual article inside the magazine by Thomas J. Willis. In urgent prose, Willis describes firefighting in the cities London and New York, and in the forests of Canada.

Irving Crump (1887–1979), who edited the magazine and was a prolific author, must have had a particular interest in firefighting because he had previously written books about the subject, namely, *The Boys' Book of Fireman* (1916) and *The Boy Scout Fire Fighters* (1917). He was later to publish a third: *Our Fireman* (1938). Clearly, he thought the subject would entertain his young male readers and, in addition, the saving life/public service ethos of firemen accorded with some of the values the Boy Scout movement was trying to inculcate. However,

126 J.B. Taylor, Front cover of *The Scout*, (1929). Illustration, 17 x 24 cms. London: Newnes & Pearson Printing Co. Ltd., for the Boy Scouts. Author's collection

the positive message was somewhat undermined by an advertisement on page 41 encouraging readers to tell their 'chums' about the magazine's 'Fire Number' illustrated by a cartoon of a boy scout about to throw a bucket of water over a sleeping fireman, which was hardly a friendly or respectful action.

## 127 — 1930

## John Player & Sons, Fire-Fighting Appliances

Cigarette cards, which in the United States date from 1879, are a sub-category of trade cards. As explained earlier, American tobacco manufacturers first issued pictorial cards in the 1880s and the practice soon spread to Britain. The pictures consisted of reproductions of photographs or specially commissioned paintings many of which were themselves based on documentary photographs. Often, those responsible for the artwork are not credited on the cards although experts know the names of some.

Cigarette cards were devised precisely in order to be collected (the habit is called 'cartophily') and were issued in sets to encourage brand loyalty. They were and are saved and traded by both adults and children. Besides the pictures themselves, printed information on the backs of cards endowed them with educational value; consequently, a set of cards constituted a miniature encyclopaedia. The subject matter of the cards included famous actresses, clowns, kings and queens, politicians, sporting heroes, the Boer War, the British Empire, animals, automobiles, flags, flowers, ships, Turner's oil paintings and fire engines.

The ten British cards illustrated here date from 1930 and were issued by John Player & Sons, a division of the Imperial Tobacco Company (ITC), Bristol, which was founded in 1901. Entitled *Fire-Fighting Appliances*, the complete set comprised 50 cards. As the illustration demonstrates, a proportion of the cards included portraits of early firemen and twentieth-century brigades tackling blazes. In fact, the set provided a potted history of firefighting mainly in Britain. ITC commissioned Mardon, Son & Hall Ltd, a design, packaging and colour-printing firm located in Bristol, to produce its cigarette cards and so it was probably one or more of the 15 full-time illustrators employed in its 'Artists Department' that created the appliance set.

Since thousands of sets and millions of cigarette cards were produced between the 1880s and 1940s, and had a wide currency among the population, they must have helped to spread knowledge about the subjects depicted everywhere. Historic sets in good condition are now valuable collectibles; however, it is possible to buy fac-

127 John Player & Sons, *Fire-Fighting Appliances*, (1930). Ten cigarette cards. Reproduction courtesy of Imperial Tobacco Ltd., Bristol. Author's collection

128 Haig Patigian, *Monument to the Volunteer Fire Department of San Francisco*, (1933). Memorial/public sculpture, bronze, over life-size. San Francisco, CA: Washington Square Park. Photo © John Dugger, San Francisco

similes of sets, such as the appliances one, from specialist suppliers. ITC have granted Murray Cards (International) Ltd of Hendon, London the right to reprint its cigarette card series. A substantial collection of cigarette cards assembled by Edward Wharton-Tigar is held by the British Museum.

## 128 — 1933

### Haig Patigian, Monument to the Volunteer Fire Department of San Francisco

As explained earlier, when firebelle 'Lillie' Hitchcock-Coit died in 1929, she bequeathed part of her fortune for the erection of a memorial to San Francisco's Volunteer Fire Department (1849–66). It took the form of a bronze, over life-size, group statue of three firemen, two of whom are wearing fire helmets, tunics and boots. One of them gestures skywards with his right arm and carries a speaking trumpet in his left hand; the second fireman crouches to control a snake-like hosepipe; the third figure, bareheaded and dressed in oilskin and boots, cradles a swooning female in his arms. The figures are artfully arranged on a circular, white marble base so that the sculpture works 'in the round'. It was unveiled at a dedication ceremony held on 3 December 1933 attended by local dignitaries and firemen and is situated in a small park called Washington Square.

The artist responsible for the memorial was Haig Patigian (1876–1950), a man of Armenian origin who came to the United States in 1891. Patigian was apprenticed for a time to a sign painter and then studied at the Mark Hopkins Art Institute in San Francisco, the city that henceforward became his home. (Later, he also spent two years in Paris receiving tuition from Alix Marquet.) Initially, he made a living producing cartoons and illustrations but then he turned to sculpture and became successful fulfilling commissions for statues of American Presidents and allegorical subjects such as Creation. Patigian was opposed to modernism and the left-wing, social realist art of the 1930s and so it is not surprising his firemen's memorial continues the nineteenth-century tradition of classical-style public sculpture, whose figures combined naturalism and gestural rhetoric.

## 129 — 1933

### Unknown photographer, Ed Wynn as The Fire Chief

Isaiah Edwin Leopold (1886–1966), stage name 'Ed Wynn' aka 'The Perfect Fool', was a popular vaudeville and Broadway comedian or clown who wrote and produced his own shows. His comic trademarks were: dishevelled hair, rubbery face, big round spectacles, silly hats, oversized shoes, badly fitting clothes, squeaky voices, lisping speech, outlandish puns, fluttering hands and absurd inventions. Eventually, he appeared on radio and

129 Unknown photographer, *Ed Wynn as The Fire Chief*, (c. 1933). Publicity still. Author's collection

130 Bill Holman, *Smokey Stover*, (1935). Newspaper comic strip distributed by *Chicago Tribune – New York Daily News* Syndicate. © 1935 Chicago Tribune – N.Y. News Syndicate, Inc

television, and acted in several serious dramatic films. From 1932 to 1935, he starred in a live, weekly radio show called *The Fire Chief* (NBC), which was sponsored by Texaco (now Chevron Texaco Corp.), a company founded in 1902, in order to promote its new 'Fire Chief' brand of gasoline.

Various kinds of merchandise exploited the Fire Chief character; for example, an illustrated book for boys and girls entitled *Fire Chief Ed Wynn and his Old Fire Horse* (1934), a chalk-ware portrait bust, a game, a jigsaw puzzle, a siren pipe, and a 10–cms-high, jointed wooden figure toy. Collectors now covet such 1930s' memorabilia. Items associated with Texaco are also preserved in Trump's Texaco Museum, Knightstown, Indiana.

In 1933, Metro Goldwyn Mayer sought to capitalise on Wynn's success by releasing a feature film about the Texaco Fire Chief. *The Chief* (1933, aka *My Old Man's a Fireman*), directed by Charles Reisner, is black-and-white and runs for 65 minutes. It is set in the Bowery, New York during the 1890s. Wynn plays Henry Summers, a bumbling, timid, honorary Fire Chief who runs for political office against a gangster. Henry's father had been a firefighter who died saving others and in the first scene, the son unveils a statue in his memory but when he turns on a fountain, he drenches the watching crowd. The movie continues in the same farcical manner until the likable fool himself becomes a hero and triumphs over his corrupt rivals. As can be seen in the publicity photograph, Wynn wears a fire hat but it is ludicrous because it is far too small.

Another publicity photograph shows Wynn at the wheel of a fire engine, wearing a high-fronted fire helmet, and with a cigar in his mouth.

During the nineteenth century, as we have seen, fire chiefs – apart from the fictional Darktown one – were portrayed as admirable and heroic figures even by cartoonists. Gradually, however, they became figures of fun. Of course, they were authority figures against whom actors such as Charlie Chaplin playing ordinary firemen needed to rebel.

The change in perception may also have been because some fire chiefs were incompetent while others – such as 'Boss' Tweed – used the position as a route to political power and so were associated with sleaze in local government that Henry Summers, ironically, was fighting.

Decades later, Hollywood screen comedian Steve Martin played Fire Chief Charlie D. Bales in the romantic comedy *Roxanne* (1987), which was based on Edmond Rostand's play *Cyrano de Bergerac* (1897). Bales is a cultured, intelligent, articulate, witty and sensitive man who is cursed with an enormous nose, which makes him aggressive towards those who make fun of it. He manages a small town firehouse with a bumbling crew whom he trains in eccentric ways such as hose drills accompanied by waltz music.

## 130 — 1935

### Bill Holman, Smokey Stover

Since firefighters seem to have appeared in every visual medium imaginable, it is no surprise they can be found in newspaper comic strips.

The latter emerged in the 1890s and were developed during the early twentieth century. Bill Holman (b. 1903, Crawfordsville, Indiana; d. 1987) was a prolific American cartoonist with a talent for verbal and visual puns, and screwball comedy. During the 1930s, he invented the character Smokey Stover, who was a fireman by occupation. Smokey made his debut in a comic strip published in the 'Funnies', a Sunday supplement of the *Chicago Tribune*, on 10 March 1935 and he proved so popular the strip was syndicated to other papers and continued until 1973. During World War II, American aircrews painted images of Smokey on their bombers. Anthologies of the strip, which were published in book form in the 1940s, are now collectors' items. Additional merchandise included Tru-Vue, 3D, Viewmaster, 35 mm films. In May 2001, a small statue of Smokey (No. 21 in the Classic Comic Character series) was issued in a limited edition by Dark Horse Comics Inc., of Milwaukie, Oregon.

Smokey is a slim, hyperactive individual, fond of pranks and prone to gaffs. (He has been dubbed 'the nuttiest fireman in comics' history'.) His normally startled eyes are huge and his round face lacks a chin. He dresses in a red shirt, black trousers and grey boots, and wears his helmet backwards so that its neck guard looks like a beak. Smokey's boss is Fire Chief Cash U. Nutt, whom Smokey sometimes causes to leap from his trousers to reveal red-flannel long johns. Smokey often drives an egg-shaped, two-wheeled car known as 'the foo mobile' and he is called 'the foolish foo fighter'. 'Foo' was a nonsense word Holman employed and claimed to have discovered on a Chinese figurine. It was widely used in the English-speaking world during the 1930s and 1940s.

Holman's strips were coloured red, yellow and blue, but the hues were rather pale because of the absorbent newsprint they were printed on. As the story about the fish cooked in a fire Smokey helps to extinguish, Holman's humour was of a gentle kind that appealed to children and adults alike. In 1961, Holman was honoured by being appointed president of the National Cartoonists Society, a professional organisation founded by American cartoonists in New York in 1946.

## 131 — 1935

### Unknown writer and illustrators, The Picture Book of Firemen

Published by Ward, Lock & Co., Ltd of London and Melbourne around 1935, *The Picture Book of Firemen* (one of the Opal Picture Books series) is representative of a considerable body of non-fiction educational literature aimed primarily at boys. Two earlier examples were: Frank Mundell's *Stories of the Fire Brigade* (London: Sunday School Union, 1895) and *Story of the American Firemen* (New York: McLoughlin Bros, 1908). A later example was: Lisa Peters' *The Wonder Book of Firemen and Fire Engines*, (New York: Grosset & Dunlap, 1956). All three books were illustrated. The anonymously written text consists of a question and answer dialogue between a father and his son Bobby (perhaps aged between five and ten) all

131 Unknown writer and illustrators, *The Picture Book of Firemen*, (c. 1935). Cover of children's book, 18 x 25 cms. Author's collection

about the work of firemen, the machines and apparatus they use, and the techniques they employ. Bobby also tours a fire station and so is able to question the professionals directly.

To increase the child's involvement, there are images of boys wearing brass helmets and holding hoses, and riding toy fire engines. There are also images of a girl's fire brigade practising and boys' brigades drilling at Dr Barnardo's (a home for orphans) and Eton College (a public school). Already imbued with gender prejudice, Bobby is contemptuous of girls as firefighters but his more enlightened father defends them and cites as evidence of their abilities what women had already achieved in the realms of aviation and sport.

Most of the book is about Britain but there is a final chapter on firefighting in other countries. It is generously illustrated with colour images and monochrome documentary photographs plus some film stills. However, many of the colour images are unpleasantly lurid because they are photographs that have been painted and airbrushed over.

Exceptionally, two colour illustrations do feature the signatures of artists. The image on the cover, by an uncredited artist, is clearly a reproduction of a painting and the axe-wielding fireman at the top of a ladder is dressed in the classic uniform and brass helmet established by Captain Shaw in Victorian times. Indeed, the subject and composition is very similar to a chromolithographic print of a fireman, one of a series depicting trades and professions produced during the nineteenth century (see illus 67).

This picture book was clearly an attractive and informative gift and no doubt encouraged boys to play at being firemen. It increased the boy reader's knowledge of the profession and may well have planted the idea that one day he too could become a fireman.

## 132 — 1936

### Reg Speller, LCC Firemen or London Firemen change Helmets

This amusing photograph, taken in London in early 1936 by Reg Speller a photojournalist working for Hulton Press, can be regarded as a tribute to the endur-

132 Reg Speller, *LCC Firemen or London Firemen change Helmets*, (1936). Photo source: Getty Images, Inc., Hulton Archive

ing appeal and importance of firemen's helmets. It depicts a moment of transition when the famous brass helmet, the Merryweather 'Metropolitan' (on the right) of the London Fire Brigade, introduced in 1866 by Captain Shaw, was replaced by a more streamlined black cork or composition one (on the left). The reason for the change was that brass helmets conducted too much heat and electricity. (Firemen could be electrocuted if live electric wires touched their metal helmets.) The cork model – a 'super safety helmet' designed by Merryweather & Sons in 1934 – was lighter and more heat resistant.

Since the two firemen in the centre of the composition are seen in profile, their silhouettes are virtually identical and resemble a mirror image. Viewers are able to compare and contrast the shapes of the old and new helmets with ease. It is clear from the strange, intimate, stomach-to-stomach pose of the central figures that this was a scene specially staged for the camera. (The two men seem to struggling to keep straight faces.) One may conclude it was a publicity shoot, and that the final image was intended to be a close-up of the two heads. However, because this photograph also shows two assistants holding a white rectangular sheet of paper to provide a plain background for the headshot, it serves to reveal the artificial character of the event.

133a Gilbert William Bayes, *Firefighters' Memorial*, (1936). Bas-relief, stone and bronze. Hall of Remembrance, London Fire and Emergency Planning Authority Headquarters, 8 Albert Embankment. Photo: London Fire Brigade

133b Detail of relief

## 133a/b — 1936

### Gilbert William Bayes, Firefighters' Memorial

In May 1937, an imposing, nine-storey building made from brick and Portland Stone was completed on the Albert Embankment in London. The building, which was the new headquarters of the London Fire Brigade, was designed in an art deco style by Edwin Paul Wheeler (1874/5–1944) and featured sculptures with firefighting themes by two British artists: Stanley Nicholson Babb and Gilbert William Bayes.

Babb (1873–1957), who is now little known, supplied carved Portland stone relief panels for the façade (placed over the north and south entrances) showing firemen in action with hoses and ladders. A third panel, which is now lost, depicted a female figure personifying the Fire Brigade holding a fire engine in each hand. Babb was born in Plymouth, Devon and studied in France and Italy. He produced bronze figurative sculptures and World War I memorials and exhibited at the Royal Academy from 1898 and won a gold medal in 1902.

Bayes (1872–1953), a much better-known artist, provided four, oblong, stone reliefs for the façade (placed above the central entrance and above windows), which adopted a more fanciful and mythological approach to the subject with representations of a fire-spouting Dragon, the Sun God Helios in his Chariot, Mermaids fighting fire, and two galleons. Inside the Headquarters, Bayes also supplied a large bas-relief memorial carved from stone showing in profile a team of seven, contemporary firemen tackling a blaze with hoses and ladders. One figure is wearing a breathing apparatus. Assyrian and Ancient Greek architectural friezes inspired Bayes' fondness for low relief sculptures. His method was to produce a full-scale model in clay or plaster and then employ stonecutters to rough out the relief, which he would then finish. The quality of the linear draughtsmanship in this particular relief is especially fine. The sculpture is framed in bronze, which includes panels on either listing the names of fallen firefighters while above is a Latin motto stating 'The end crowns the work'. A further bronze relief mounted on top depicts a horse-drawn fire engine travelling at full gallop. Bayes, apparently, had a passion for horses. Members of Lloyds Underwriters presented this work to LFB.

There is a second, similar stone and bronze relief by Bayes dating from 1940 in possession of the LFB, which is historical in character. The stone relief shows an early fire brigade, a horse-drawn fire engine plus salvage men in action. This time the panels on either side have full-length, bas-relief, bronze portraits of insurance company firemen and the Latin motto reads: 'The beginnings of all things are small.' The sculpture was commissioned by the insurance companies whose private fire brigades formed the London Fire Engine Establishment in 1833.

Bayes was born in London into an artistic family and studied at the Royal Academy Schools from 1896 to 1899. After being awarded a gold medal and travel grant, he spent time in Paris and Italy. He was a highly prolific, versatile and commercially successful academic artist who created bronze portrait and equestrian statues, war memorials, medals, designs for Doulton stoneware, ornamental reliefs and friezes. His particular interests were in the use of colour and sculpture for decorative purposes. A monograph and catalogue raisonné about Bayes, written by Louise Irvine and others, was published in 1998.

## 134 — 1938

### Unknown photographer, Horse-drawn steam engine used in the film In Old Chicago

Henry King directed and Darryl F. Zanuck produced the 1937–38 disaster epic *In Old Chicago* – billed as 'The Great American Motion Picture!'- which starred Alice Faye, Don Ameche and Tyrone Power. Niven Busch's story 'We, the O'Leary's', inspired the script of this black-and-white, quasi-historical melodrama.

The O'Leary's were a real family of Irish origin living

134 Unknown photographer, *Horse-drawn steam engine used in the film In Old Chicago*, (1938). © Twentieth-Century Fox

in Chicago in 1871 when a massive fire occurred. Indeed, folklore has it that Mrs O'Leary's milk cow started the fire by kicking over a lantern in a barn. Much of the film concerns the rivalry between two brothers, one an honourable lawyer and politician, and one a charming but unscrupulous gambler and property speculator, played by Ameche and Power respectively, but it merits inclusion in this book because of its lengthy, spellbinding sequence devoted to the Great Fire, which includes dramatic shots of fire companies in action. Despite some factual inaccuracies – such as the use of hand pumps – the scenes showing horse-drawn ladder trucks and steam engines racing through the streets are quite uncanny because it is as if the nineteenth-century prints of Currier & Ives and others have sprung to life. Fred Sersen was the special-effects maestro responsible for the miniatures, glass shots, travelling mattes and process screens that enabled the conflagration to be realistically portrayed on screen.

In 1871, Chicago (aka 'Queen of the West'), an inland port on Lake Michigan and railway hub, was a booming city. Most of its sidewalks and 60,000 buildings were made of wood. Around 334,000 lived there when, on the sweltering evening of 8 October, a fire started in a barn near DeKoven Street. Unfortunately, Chicago's fire companies – equipped with 16 steam fire engines – were depleted and exhausted by earlier fires and a series of mishaps and errors resulted in their late arrival. Fanned by a high wind, the fire spread relentlessly from district to district cutting a path a mile wide and overwhelming the fire department's resources despite the sustained efforts of the fire chiefs and their men. Urgent requests for help were dispatched to the fire departments of nearby cities. Crowds, unruly mobs, horses and vehicles clogged the streets and bridges as panic-stricken citizens strove to escape incineration. Lieutenant-General Philip H. Sheridan of Civil War fame, who has a cameo role in *In Old Chicago*, was in the city and authorised the use of gunpowder to blow up buildings to create firebreaks; he also imposed martial law to deter looters. By the time the fire ceased 30 hours later on the morning of 10 October, when rain fell, one third of Chicago had been destroyed and between 250 and 300 of its citizens had perished, and 70,000 were left homeless.

Naturally, the Great Fire prompted many news reports and illustrations in the press and so it quickly became internationally famous. One eyewitness survivor was the New York artist John R. Chapin (1823–1907) whose vivid illustration of crowds fleeing a wall of flame across the Randolph Street Bridge appeared in *Harper's Weekly*. Chapin had a high viewpoint from a window in the Sherman House hotel. Later he recalled feeling like Nero watching Rome burn.

London-born Alfred R. Waud (1828–91), another visual artist who worked for *Harper's* and *Every Saturday*, heard about the fire in St Louis and rushed to the scene by train. His 31 drawings about the fire were reproduced in *Every Saturday*.

Instant histories quickly appeared and later there were dioramas, commemorative poems, novels, pageants, prints, allegorical paintings, sermons, songs and stage performances with painted panoramas. In 1878, for example, the Griswold Opera House in Troy, New York State advertised a 'very realistic and sensational panoramic and dioramic drama "Chicago" ... with a grand panorama, 1600 feet long'. The poster included images of fire companies battling the blaze, crowds fleeing and vigilantes lynching an arsonist from a lamppost. In the early 1890s, promoters Isaac N. Reed and Howard H. Gross produced a cyclorama about the fire. Spectators stood in the middle of a large canvas (20,000 square feet) consisting of five interconnected scenes, which an international team of ten artists had created.

As *In Old Chicago* demonstrated, interest in the Great Fire continued into the twentieth century. Later, in June 1960, a huge entertainment park entitled Freedomland U.S.A. opened in the Bronx, New York. It was divided into seven sections each of which recreated famous historic places in America. One such place was Chicago in 1871. Visitors were enthralled by a live re-enactment of the Great Fire involving mock buildings made from steel-and-asbestos, real flames, a firehouse, manual pumps and firemen directing hoses. Spectator participation was encouraged because volunteers were invited to help operate the pumps.

This extraordinary simulation ceased in September 1964 when Freedomland closed. An earlier fire and firefighting simulation took place at an entertainment park called Wonderland (1906–11), Revere Beach near Boston, Massachusetts. In 1971, on the hundredth anniversary of the fire, documentaries about it were screened on American television and firefighting demonstrations were held and vintage steam engines paraded on Chicago's Soldier Field stadium. Today, of course, there are websites devoted to the history of Chicago's Great Fire.

## 135 — 1938 or 1939

### Frank Newbould, AFS, London needs Auxiliary Firemen now

In 1938, anticipating war against Germany and the bombing of Britain's cities, the nation's Civil Defence authorities decided to drastically expand the fire service by recruiting civilians (89,000 men and 6000 women; 23,000 in London) and training them for fire duties to augment the work of the regular fire brigades. This body of men and women was named the Auxiliary Fire Service (AFS). War leader Winston Churchill was later to

135 Frank Newbould, *AFS, London needs Auxiliary Firemen now*, (1938 or 1939). Lithographic poster, 34.5 x 53 cms. London: H.M. Stationery Office. Author's collection

dub them 'heroes with grimy faces'. In images, they can be readily distinguished from regular firefighters because they wore a different kind of helmet – a dome shape made from grey steel with a circular brim and chin strap, a military type worn by British soldiers. The fireman's double-breasted tunic was navy coloured and had shiny metal buttons; spanners for opening hydrants and axes for ventilating buildings were suspended on a thick belt and oilskin over trousers and rubber boots protected his legs from water. Uniforms for women varied because 20 different kinds were issued during the war. In emergencies, firefighters also had to don anti-gas masks and protective clothing.

Posters, of course, are a centuries old, image-text method of communication and recruiting ones for war and for the professions had been popular since the time of World War I. Frank Parkinson Newbould (1887–1951), the designer of the AFS poster, was born in Bradford, Yorkshire and studied at Bradford and Camberwell colleges of art. During the inter-war period, he became noted for rail travel and sports posters, and for book illustrations. Newbould employs the familiar hosepipe motif in which two firemen grapple with a thick, heavy pipe and nozzle. Since in this case the men are seen from behind, the composition invites the viewer to lend a helping hand. How to join in is indicated by the instruction printed at the bottom of the poster: 'Enrol at any fire station.' The figures are rendered in a modern, simplified, flat style that Henri Toulouse-Lautrec perfected in his lithographic posters, influenced by Japanese prints, during the 1890s. It suits the graphic medium and provides strong, dark silhouettes against an orange-red background (representing a wall of flame), which has a mist-like appearance due to the pigment being sprayed on with an airbrush.

In 1942, Newbould joined the War Office where he assisted the well-known designer Abram Games.

## 136 — 1939

### Felix H. Man, An Officer of the Auxiliary Fire Services Prepares for the Worst

On 27 May 1939, the British weekly, illustrated magazine *Picture Post* published a four-page article entitled 'The Auxiliary Fire Service', which was No. 3 in a series called

136 Felix H. Man, *An Officer of the Auxiliary Fire Services Prepares for the Worst*, (1939). Photograph on front cover of *Picture Post*, Vol. 3 No. 8, 27 May 1939. Author's collection

'Britain Prepares'. The magazine's cover featured a large, impressive photograph of the head-and-shoulders of a uniformed auxiliary fireman gazing warily towards the sky. The circular shape of his helmet becomes a halo around the man's head and the metal buttons and AFS badge on his tunic are clearly visible. Inside were 13 black-and-white documentary photographs of AFS personnel, both men and women, drilling and rehearsing for the onslaught to come. The accompanying short text ends by saying that the AFS needs more men and women between the ages of 25 and 50, and so the whole article served a recruiting function. One advantage for male volunteers was that firefighting was a reserved occupation and so they were excused military service.

*Picture Post* (Hulton Press, 1938–57), a hugely popular and influential magazine with 'a social eye', was read by an estimated five million. Stefan Lorant (1901–97), the magazine's first editor decided its proportions should be tailored to those of photographs taken with 35mm cameras.

Lorant, a Hungarian of Jewish origin who had been arrested and deported by the Nazis in 1933, was a pioneer of photojournalism, that is, journalism in which a sequence of photographs told a story and dominated the text. The development of photojournalism using available light to capture 'fruitful moments' was facilitated by the arrival of small, hand-held cameras such as the Ermanox and Leica. Working closely with editors, the photographers not only supplied the images, they researched and wrote the articles.

The uncredited photographer of the AFS images was Felix H. Man (1893–1985, aka Hans Felix Sigismund Baumann) who was born in Freiburg, Germany. He was multitalented: an artist, photographer, writer and collector of artists' lithographs. He had been given his first camera – a Kodak – at the age of nine. Before and after World War I he studied fine art and the history of art in Berlin and Munich. By the late 1920s, he was making drawings and taking photographs for picture agencies and for daily newspapers published in Berlin by Ullstein. Man arrived in England in 1934 a year after Hitler came to power and from 1938 to 1945 he was *Picture Post*'s chief picture correspondent. It was ironic that the photojournalist responsible for the article about Britain preparing to resist German bombing should himself have been a German. Later, in 1948, Man became a British citizen.

On 1 February 1941, *Picture Post* updated the AFS story by publishing a seven-page, picture-dominated article entitled 'Fire-fighters!' The cover photo was a head-on shot of two firemen gripping a nozzle spurting water. Inside, there were shots of men and women calmly operating telephones in London's Central Control Room. These were accompanied by more dramatic photographs – some full page – of roof spotters identifying the first fires, firemen sliding down poles and manning engines, a dual purpose pump leaving a fire station, firemen at work in the streets, on roofs and at the top of a long turntable ladder. Captions written in an urgent, telegraphic style emphasised the courage, hard work and heroism of the firemen and the risks they had to take.

Bert Hardy (1913–95), a working class Londoner who was the first photographer to be credited by name by *Picture Post*, took these images during the Blitz with a Leica. A sub-heading explained that he had slept every night in a fire station for a fortnight waiting for an incident. Then came a severe bombing raid and he spent all night with the firemen during which time he lost a camera and tripod worth £50 and a pair of trousers due to his leg being burned.

A further issue of *Picture Post* – 3 May 1941 – contained an article entitled 'The morning after the Blitz', which documented the impact of the bombing on London's buildings and civilians whose homes had been hit. Firemen were photographed damping down smouldering ruins, retrieving their equipment and clearing up.

137 Leslie Gilbert Illingworth, *Auxiliary Fireman*, (1940). Drawing, pencil, brush and Indian ink, 28 x 18 cms, published as a newspaper cartoon. Photo courtesy of Department of Pictures & Maps, National Library of Wales, Aberystwyth

## 137 — 1940

### Leslie Gilbert Illingworth, Auxiliary Fireman

On 7 September 1940, the inexperienced AFS was tested to the limit when 600 Luftwaffe aircraft bombed the East End of London from dusk to dawn. The attacks continued for 57 consecutive nights. On 12 September, the *Daily Mail* newspaper published a powerful cartoon depicting a full-length portrait of a robust auxiliary fireman holding a hosepipe issuing a jet of water. The man has a resolute expression on his face and stands calmly amidst the debris of a bombsite while in the background a nurse tends to the injured and a fire warden comforts a weeping female civilian. Above them, the sky is filled with clouds of black smoke. This brush-and-ink drawing, executed with considerable verve, was clearly intended as a tribute to the courage and effectiveness of the AFS. The artist responsible was Leslie Gilbert Illingworth (1902–79), a prolific cartoonist and book illustrator, who was born in Wales and studied at Cardiff Art School and at the Slade School of Art in London. By 1940, he was working as chief political cartoonist for the *Daily Mail* newspaper.

At first, AFS firefighters had a somewhat negative image in the public's mind because, during the phoney war period, they appeared idle: they were unfairly called 'army dodgers' and 'the darts and snooker brigade'. However, after their courageous efforts and sacrifices during the Blitz, their popularity was assured.

## 138 — 1940

### Leonard Rosoman, A House Collapsing on two Fireman Shoe Lane, London

Among the AFS recruits were some from the professional class who included poets such as Stephen Spender and painters such Leonard Henry Rosoman (b.1913). Rosoman has had a long career as a traditional figurative artist, book illustrator and art school tutor. During the early 1930s, he studied fine art at the King Edward VII Art Department of Durham University and at the Royal Academy Schools and the Central School of Art, in London. He was later elected a member of the Royal Academy. Before he was mobilised in September 1939, he had been a part-time member of the fire brigade.

His dramatic oil painting *A House Collapsing on two Firemen, Shoe Lane, London EC4* (1940) captures a moment of extreme danger. It is tantalising because it is like a film with a cliffhanger ending: one wonders, did the men threatened by the falling wall manage to escape or were they killed? In fact, the painting was based on a real incident Rosoman observed while on duty. He narrowly escaped death or injury by ducking into a nearby doorway but two of his colleagues were not so fortunate. The principal colours are red, yellow, beige and black. The latter is the hue of the falling wall, which cuts across the composition in an emphatic diagonal. Bricks fly into the air as the top of the wall crashes into a row of houses on the right. Another of Rosoman's wartime oils, painted in 1943, depicts the twisted wreckage of a burnt-out 'Heavy Unit' beneath a gloomy sky. (A 'Heavy Unit' was a fire engine with pump and ladders devoid of brass and copper elements, which was painted battleship grey.) This work closely resembles one of Graham Sutherland's wartime paintings. Its neo-romantic style was appropriate to the times because members of that particular British art movement were nostalgic about ruins, which, of course, the war supplied in abundance. Rosoman told a radio interviewer – *Desert Island Discs*, (BBC Radio 4, 9 June 2002) – that he preferred *A Burnt-out Fire Appliance* to *A House Collapsing* because it was not so sentimental and melodramatic. Both paintings are in the collection of the Imperial War Museum, London.

In 1942, Rosoman was commissioned by the Home Office to illustrate a new Fire Service Manual. For nearly two years, he travelled the country making technical drawings of firefighting equipment. After serving in the AFS, Rosoman was appointed Official War Artist to the Admiralty and posted to the British Fleet in the Pacific.

138 Leonard Rosoman, *A House Collapsing on two Firemen, Shoe Lane, London EC4*, (1940). Painting, oil on canvas, 91.4 x 76.2 cms. London: Imperial War Museum. Reproduced courtesy of the artist

139 Achille Beltrame, *Firemen in London*, (1941). Illustration published in *La Domenica del Corriere*, 19 January 1941. Photo courtesy of the Mary Evans Picture Library, Blackheath

## 139 — 1941
### Achille Beltrame, Firemen in London

This colour illustration can be dubbed 'the ultimate hose shot' because it shows a long line of AFS firemen directing their hoses in unison onto a blaze caused by nocturnal Nazi bombers. The image is striking because of the sheer number of firemen depicted and the repetition of hoses and water jets. One doubts the London brigades ever employed so many men in a single action and had such a superabundance of water. What the image resembles, in fact, are photographs of peacetime hose drills in which rows of fireman performing similar tasks appear.

Beltrame (1871–1945), the artist responsible, was a prolific and well-known Italian painter, poster artist and illustrator whose images of current events regularly appeared on the cover of *La Domenica del Corriere* (Sunday Courier, a weekly founded in 1899). Since Beltrame was not an eyewitness to the events he depicted, he must have relied on photographs and his imagination. His style, a popular realist one, was formed in the nineteenth century and changed little over the years. When his image was published, on 19 January 1941, Britain and Italy were at war and in North Africa, the British army was defeating Italian forces. Although Beltrame's image is primarily descriptive, one senses he admired the strength and organisation of the London firefighters.

## 140 — 1939
### Barribal, Duty Calls

When women began to join the Auxiliary Fire Service just before the outbreak of World War II, it prompted images of firewomen in popular media such as postcards. For instance, a head-and-shoulders portrait of an AFS woman seen in profile appeared on a postcard published by the Dundee Company Valentine & Sons. She wears lipstick, a peaked cap and displays a red AFS badge on her dark blue jacket. The card's title was *Duty Calls* and the artist was 'Barribal' (that is, William Henry Barribal,

140 Barribal, *Duty Calls*, (1939). Colour picture postcard, 14 x 9 cms. Dundee & London, Valentine & Sons Ltd. Author's collection

141 George Gibbons, *Women! You are needed in the National Fire Service*, (1941). Recruiting Poster for the National Fire Service issued by Her Majesty's Stationery Office, 74.3 x 50.2 cms. London: Imperial War Museum

1873–1956), an English painter and popular illustrator of postcards and playing cards, whose speciality was watercolour or gouache images of cute children and upper-class Edwardian beauties. Barribal's glamorous representation suggested that even elegant, fashion-conscious women were willing to answer the call of duty and would do so in style.

## 141 1941

### George Gibbons, Women! You are needed in the National Fire Service

In August 1941, the AFS and the regular fire brigades were combined into a single National Fire Service (NFS). This poster, by George Gibbons a little-known designer, published by HM Stationery Office was aimed specifically at women and a caption told them what jobs were available: 'telephonist, despatch rider, driver, canteen worker and for many other duties.' Although some recruiting posters for the AFS had used photographs of particular individuals, in this case, the designer has settled on an idealised portrait head, unnaturally coloured red and green, set against a blue background. In fact, the red area that encompasses her chin, mouth and nose could be interpreted as the reflected light from a fire. The attractive but not excessively beautiful young woman gazes into the distance and appears self-assured.

Some firewomen were trained in the use of pumps and a few did fight fires but for the most part their role was supportive. Nevertheless, they were exposed to danger because observation posts and fire stations with their watch or control rooms were bombed. In addition, mobile kitchen staff, the drivers of canteen vans, lorries and staff cars, motorbike despatch riders and field telephone units were exposed to air raids, machine gunning and accidents in the streets. Miss Gillian 'Bobby' Tanner was the only AFS firewoman to be awarded the George Medal for her courage and coolness in circumstances of extreme danger. On 20 September 1940, she drove a 30 cwt. lorry loaded with 150 gallons of petrol through London's streets to the docks and then distributed the petrol to fire pumps for three hours during intense bombing. An image of her in action is preserved in the museum of the London Fire Brigade: Mary Pitcairn's painting *Driving by Moonlight* (1941), which is a close up of Tanner at the wheel of her lorry as viewed though the windscreen. (A similar image of a woman driver can be found in John Cosmo Clark's [aka Cosmo Clark RA, 1897–1967] recruiting poster *Women Required for Motor Driving & Telephonist Duties.*) Alfred Reginald Thomson (1894–1979) also painted a three-quarter-length portrait of Tanner standing in the canteen of a fire station, which is now in the collection of the Imperial War Museum.

In total, 23 women died in action and their names are recorded on a bronze panel at the base of John W. Mills' 1991 bronze memorial to firefighters who were killed, entitled *Blitz*, situated near St Paul's Cathedral (see illus 181). Their names are flanked by relief depictions of two figures – a Despatch Rider and an Incident Recorder.

The recruitment of women for the AFS and the NFS was the first time in history that several thousand had been given the opportunity to join the fire service and they acquitted themselves well. However, in 1945, they were demobilised and it was not until 1982 that women were permitted to join Britain's regular fire brigades.

## 142 1944

### Reginald Mills, front cover of "In the Service of the Nation": The N.F.S Goes into Action

## 143 1944

### Reginald Mills, Control Room in Action

Before the war, some members of the AFS and the NFS had been professional fine artists or art students or commercial, furniture and theatre designers, and illustrators, and they continued to create in their spare time between alarms and when off duty. (Normally, they worked 48 hours on and 24 hours off.) They included: Leslie Baker, Paul Dessau, Stanley Flegg, Wilfred Stanley Haines, Bernard Hailstone, Rudolf Haybrook, Norman Hepple, Reginald Mills, Francis Needham, Philip North-Taylor, Leonard Rosoman, Fredk J. Smith, J. Kingsley Sutton, E. Boye Uden and W. Matvyn Wright. Haybrook, a theatre designer in peacetime, was one of the principal organisers of the group. From 31 May to 5 June 1940, he was aboard the London fireboat *Massey Shaw* (1935–71) when it sailed to Dunkirk three times to rescue hundreds of British troops from the beaches and he later made a pictorial record of what he had witnessed. There were also volunteer firefighters, such as T. Butler, who were or who became amateur artists.

At least five women artists participated, including Enid Abrahams who had studied at the Royal Academy Schools and Mary Pitcairn, a painter and novelist, who was honorary secretary of the Firemen Artists' Committee.

The experience of fighting fires under night skies illuminated by explosions, flames, flares and searchlight beams and the scenes of urban devastation they witnessed gave the artists plenty of new subjects. Hepple, in a painting of St Paul's seen from Cannon St, in which a fire crew direct water at leaping flames clearly regarded the latter as aesthetically beautiful. Other artists found it therapeutic to draw and paint the traumatic incidents seared on their minds during their periods on duty. One artist – Dessau, who had studied at St Martin's School of Art – even depicted a nightmare he had experienced in which he was battling the flames but everything that could go wrong with his equipment had gone wrong. Dessau also painted girders twisted by heat, colleagues extinguishing incendiaries with buckets and a half-length portrait of Fire Force Commander Geoffrey V. Blackstone (which is now in the Imperial War Museum Collection), who was awarded the George Medal for gallantry and later wrote a book entitled *A History of the British Fire Service* (1957). 'Centurion' (that is, Ralph Tuck), described the Commander as 'tall, broad shouldered, with an alert, cheerful face'. By painting Blackstone in uniform, Dessau was adding to the tradition of portraiture that dated back to the nineteenth century when artists and photographers depicted such famous British fire chiefs as Braidwood and Shaw. Another fine portrait by Dessau (dated 16 December 1944) showing a London fireman resting and smoking after a night's toil, was reproduced in a magazine's colour advert for a brand of cigarettes. Given the harm smoke can cause firefighters, the advert's slogan – 'Craven 'A', for your throat's sake' – now seems singularly crass.

142 Reginald Mills, front cover of "*In the Service of the Nation*": *The N.F.S Goes into Action*, (1944). Booklet, author's collection

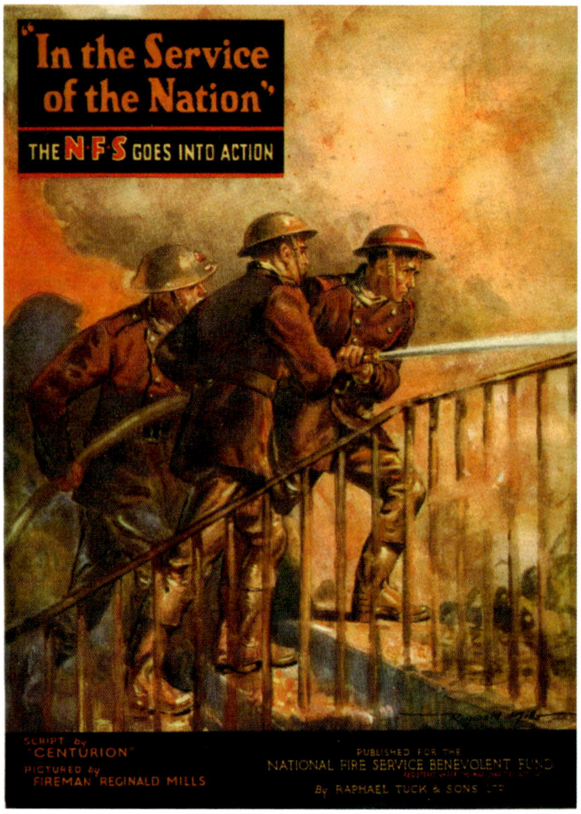


Another of Hepple's paintings – *Knightrider Street* – was actually created during an incident. Red-hot ashes rained down on the artist as he laboured and minute cinders were embedded in the pigment. Hepple (1908–94), a book illustrator and painter of academic-style portraits, had studied fine art at Goldsmiths' College of Art and the Royal Academy Schools. He was appointed Official War Artist to the National Fire Service.

Bernard Hailstone (1910–87), a book illustrator, an official war artist and painter of people and landscapes, portrayed London firewomen such as Margaret Hay, fire crews at work in rubble-strewn streets and depicted the ruins of churches and fire stations.

Fireman Reginald Mills represented hose teams in action and a relief convoy of fire appliances speeding along a country road towards a blazing city on the horizon. In addition, he depicted the less dramatic but essential tasks performed inside observation posts and control rooms where both men and women worked. In 1944, his watercolours were used to illustrate in colour a booklet entitled "*In the Service of the Nation*": *The N.F.S Goes into Action*, written by 'Centurion', which was sold to benefit the National Fire Service Benevolent Fund. Royalties from a volume published by Staples Books Ltd in 1942 – *London's Hour: As seen through the Eyes of the Fire-Fighters* – were donated to the London Fire Service Benevolent Fund. Among the reproductions inside was an unusual painting by Frederick T.W. Cook influenced by the surrealist Salvador Dalí entitled *Today's Portrait*, which showed a disembodied head wearing a respirator situated in a landscape setting. Most firemen-artists ignored the formal experimentation and non-realist imagery of modern art and produced figurative works in a naturalistic style because their priorities were content and reportage.

Amateur and professional artists were bonded by their common experience of fire service duty and so were willing to exhibit together. This is how the term 'Firemen Artists', which anticipated today's genre of firefighter art, came into being. During the war, several exhibitions of paintings by such artists were held in London at Burlington House/the Royal Academy, the Central School, the Leicester Galleries and the Royal Exchange. A show the RA mounted in August 1941 was opened by the Labour M.P. Herbert Morrison, then Home Secretary, and its poster featured a caricature by 'Maroc' (that is, Robert S.E. Coram) of Hitler wearing a painter's smock with the headline 'Adolf was a painter too ... come and see the art of those who stopped him painting our towns red'.

Touring exhibitions were despatched around the English provinces and, in 1942, a show toured Canada and the United States where it served a propaganda function. (The United States had recently joined the war following the Japanese attack on Pearl Harbour in December 1941.) A 50-page, illustrated paperback pub-

143 Reginald Mills, *Control Room in Action*, (1944). Painting, watercolour. Illustration inside the booklet

lished by Garden City Publishing in New York preceded the exhibition. The latter featured works by Ben Betts, Brian Gilks, Wilfred Stanley Haines, Norman Hepple, H.N. Manwaring, Philip North-Taylor, James Orr, Leonard Rosoman and E.A. Turpin, and was accompanied by a spokesperson, fireman-artist W. Matvyn Wright who had studied with Rosoman at the King Edward VII School of Art. Manchester Art Gallery owns one of Wright's oil paintings entitled *Firemen on a Roof* (1941), which depicts a hose team in action near St Paul's Cathedral. In 1944, Turpin painted his own rescue after he had been overcome by smoke and covered with debris in a basement.

Late in the war, bombing raids by the Luftwaffe decreased in number, but Britain was subject to attacks by new unmanned terror weapons such as the V–1 flying bombs and the V–2 rockets.

Ben Betts was a fireman-cartoonist who used humour to make light of some of the problems AFS firemen experienced. For example, one cartoon shows firemen struggling with, and complaining about, wet uniforms. This problem was exacerbated by the fact that AFS recruits were issued with only one uniform and therefore had no dry clothes to change into after being drenched on duty. Betts' solution: he depicts a fireman donning a swimming costume. In 1941, a 32–page booklet of Betts' cartoons was published with the title *Heroes with Grimy Faces ... Cartoons of Life in the Fire Service and the Lighter side of the Blitz*.

On 2 March 1942, *Life* magazine published a three-page article illustrated by documentary photographs and colour reproductions of some of the paintings in the US exhibition. Hepple depicted himself as one of three fireman avoiding flames or an explosion by taking cover behind a trailer pump. Next to the reproduction was a photograph of Hepple operating the same pump that had previously appeared in Man's *Picture Post* article. Firemen-artists frequently depicted the trailer pumps they operated and relied upon to charge hoses with water. Initially, in London, 2000 Austin taxicabs were commandeered to tow them. 'Godiva' trailer pumps were light but strongly built by Coventry-Climax Ltd. They had a small turning circle and so were highly manoeuvrable. Once detached from the towing vehicle, they could be manhandled by three or four firefighters even over the rubble from collapsed buildings. Dessau depicted such a scene near St Paul's in his oil painting *Sunday, 29th December* (1940). On that particular night, Nazi raiders dropped 100,000 incendiaries on the City of London.

North-Taylor, who had once been a student at Harrow, the north London public school, showed the school's Speech Room on fire after a raid in October 1940. His picture included a fireman manning a hosepipe mounted on top of a tall escape ladder overlooking flames emerging from the roof. The accompanying caption was blunt: 'Firemen on this job sometimes are burned like marshmallows on sticks.'

A photograph of fireman-artist Brian Gilks was reproduced showing him rescuing a colleague by using the traditional body-over-the-shoulders firefighter's 'lift' or 'carry'. The 'rescue' or 'saved' motif, which dates back to the 1850s, is one of the most familiar in firefighting iconography and there are several examples of images by firemen-artists depicting scenes of injured civilians and firefighters being assisted.

Through the book, exhibition and magazine article, the American public was informed about the devastation wrought by Nazi bombs in 1940 and the battle fought by the AFS to suppress the ensuing fires. (The AFS also had to cope with many accidental fires.) Of course, a blood price was paid: by the end of war, 327 firemen and women serving in London had been killed in action and over 3000 had been injured; the fabric of the city itself was rent with huge areas of ruins. Fireman-artist and tapestry designer Haines was one of the fallen. His death from a flying bomb in 1944 was recorded by Coram in a painting entitled *Time, Gentlemen Please*.

144 *Jacko with Hosepipe*, (1943). Film still from Humphrey Jennings & others, *Fires were Started*, (1943). Photo courtesy of the British Film Institute

## 144 — 1943

**Jacko with Hosepipe**

## 145 — 1941

**Bert Hardy (?), AFS Fireman on a Ladder with Hosepipe**

## 146 — 1941

**David Parker (?), AFS personnel queuing for Cocoa after an inspection in London by Clement Attlee**

As explained earlier, filmmakers had been shooting documentaries about firefighting since the mid–1890s. One of the most famous and finest films (now available on video) made about the subject during World War II was a drama-documentary entitled *Fires were Started* (1943, aka *I was a Fireman*). The Crown Film Unit, with the cooperation of the National Fire Service, produced the film and Humphrey Jennings (1907–50), a Cambridge-educated painter, poet, filmmaker, English patriot and socialist, researched, wrote and directed it. Jennings was a key figure in the 1930s' British documentary film movement, but he was also interested in surrealism and reviewers have detected its influence in certain scenes. The film was shot between February and October 1942 and recreated the efforts of mainly AFS men and women to tackle the massive fires caused in London's Docklands by high explosive and incendiary bombs dropped by the Luftwaffe during winter and spring of 1940–41. Exteriors were filmed at St Katherine's Dock where fires were staged and interiors were shot at Pinewood Studios, Iver Heath, Buckinghamshire. The cast, which consisted of actual firefighters borrowed from different stations, included William Sansom (1912–76), who wrote several texts about the AFS during the war and later became a noted author, and Loris Rey (1903–62) a Scottish sculptor in peacetime. Rey was recruited from Substation 6–W in Chelsea where he had decorated a recreation room with friezes. Born and trained in Glasgow, Rey became head of the sculpture school at Leeds College of Art in 1927 and specialised in portraiture. Around 1934, he moved to London and established a studio in Chelsea.

During the film's title credits, a British sculpture is shown. It is a shot of Gilbert William Bayes' *Firefighters' Memorial*, (1936), a bas-relief depicting London's regular firemen in action (see illus 133). This reference serves two purposes: first, it indicates the tradition of professionalism and sacrifice the AFS must emulate; and second, it signals that the film is another kind of visual representation of firefighting. The narrative begins quietly with the arrival of a new AFS recruit to 'Heavy Unit 1' based at Sub-station 14–Y in Stepney. (Sub-stations were established in a variety of buildings, such as schools, because of course it was not possible to build many new fire stations.) Drills and training follow but soon there is vivid footage of fires and the methods used

145 Bert Hardy (?), *AFS Fireman on a Ladder with Hosepipe*, (1941). Photo source: *Fire over London*. Author's collection

146 David Parker (?), *AFS personnel queuing for Cocoa after an inspection in London by Clement Attlee*, (21st February 1941). Photo source: *Fire over London*, booklet

to contain them. Some actors and crew suffered burns during the fire scenes and the local fire brigade had to be called out several times when the flames got out of hand. Local people also complained because they feared Jennings' night time blazes would attract enemy bombers.

Of course, Jennings' film was propaganda for the home front and so was designed to show what ordinary people could achieve in a national crisis. Its characters overcame any class differences and displayed cheerfulness, comradeship and teamwork. 'Jacko', a Cockney played by fireman Johnny Houghton, is killed after saving an injured colleague. His 'death' was deemed necessary because the Ministry of Information wanted to impress upon the populace that sacrifice was required to win the war. The film concludes with Jacko's funeral and with a triumphant shot of a munitions ship, which the firefighters had struggled to protect, departing on the morning tide.

When a renamed and shortened version of the film was released in April 1943, it was highly praised by cinemagoers and film critics. Indeed, it received better reviews than a similar fictional feature film about the AFS made by Ealing Studios and released in the same month. Entitled *The Bells Go Down*, it was directed by Basil Dearden and starred the serious actors James Mason and Finlay Currie, and the comedian Tommy Trinder. This film, which is also available on video, has much footage of firemen perched on the top of extension ladders. Like Jacko, the character played by Trinder – 'Tommy Turk' – is called upon to die heroically while trying to save the life of a colleague.

Shots of fires had a strong visual impact in *Fires were Started* because of the device of *contre-jour* (dark figures set against the light of the fires). The same effect was common in still, black-and-white photographs taken at night such as those illustrating Hardy's *Picture Post* article and the booklet *Fire over London* 1940–41. William Sansom provided the latter's text and W. Matvyn Wright designed the cover. The photographs, however, were uncredited. One taken, I believe, by David Parker on 21 February 1941 after an AFS inspection by Labour politician Clement Attlee, then Lord Privy Seal, is a remarkable example of group portraiture. It is taken from a high vantage point and so shows a sea of faces – dozens of firefighters standing shoulder to shoulder – looking up at the camera. Since the formalities of the inspection are over, the firefighters are relaxed, smiling and some are smoking or drinking cups of tea or cocoa. A mass effect is created but it contrasts sharply with German photographs depicting rigid ranks of marching storm troopers.

## 147 — 1943

### Gordon Parks, Men from Firehouse Station No. 4, Washington, DC, Prepare to Answer an Alarm

White males have dominated the American firefighting profession even though daguerreotype photographs of African-American firefighters show that they were employed during the nineteenth century. When African-Americans were permitted to join the fire service, racial segregation was the norm for much of the late nineteenth and twentieth centuries. In January 1943, Gordon Parks (b. 1912) an African-American photographer on a year's fellowship with the Farm Security Administration (FSA, a government agency directed by Roy Stryker), decided to document, by means of a photo-essay, the daily routine of Firehouse Station No. 4, in Washington, DC, because it was an all-black station.

Parks is a highly creative, multi-talented individual whose achievements against the odds are extraordinary. He was the youngest of 15 children in a struggling Kansas farming family. After experiencing poverty, homelessness and racial discrimination in his youth, he took up the camera in the late 1930s and was successful with some fashion shoots. By 1941, he was documenting social conditions in Chicago's slums. The camera, he

147 Gordon Parks, *Men from Firehouse Station No. 4, Washington, DC, Prepare to Answer an Alarm*, (1943). Photo source: Washington DC, Library of Congress, Prints and Photographs Division

thought, could be a weapon against bigotry, racism and exploitation. While in Chicago, Parks honed his visual awareness by visiting the Art Institute where he particularly appreciated paintings by Degas. After his period with the FSA in Washington, Parks became the first black photographer to work for *Life* magazine, dealing with a wide range of subjects, and for *Vogue*. He also spent time in Paris as a correspondent for *Life* and undertook assignments in Latin America. Parks' 1961 images of a dying Brazilian child provoked a huge public response. However, photography is by no means the only art in which he excels: he is also an accomplished composer, autobiographer, novelist, musician, painter, poet and filmmaker (he directed *Shaft* [1971]). In 1999, a large retrospective exhibition entitled *Half Past Autumn: The Art of Gordon Parks* was organised in his honour by the Corcoran Gallery of Art, Washington DC, and then travelled across America.

Firehouse Station No. 4, which was captained by J.B. Key or Keyes, had a reputation for its speed of response to alarms and had received citations for its rescue work. It had also won cups for the best decorated floats in city parades. Parks, who was just learning his craft under Stryker's supervision, set out to record every aspect of the Firehouse: he photographed the two-storey building from the outside and the clothes rack, information boards and equipment inside and, of course, the daily lives of the firemen. He showed probationers taking exercise, learning how to use rescue apparatus and listening to lectures from Lieutenant Mills about water pressure. He depicted men sleeping in their bunks and filling in time between alarms with such activities as studying, reading newspapers, woodcarving, repairing old toys for children, playing cards and chess. Thus, Parks' photographs now have historical and sociological value as a record of a particular way of life during the 1940s.

The photograph reproduced – an action scene set just inside the entrance to the Firehouse – was taken from a very low viewpoint. It depicts a ladder truck and a fire engine with seven firemen responding to an alarm: one descends a pole, while others hurriedly don boots, helmets and protective clothing. Such preparations only took a few seconds. We know that Parks accompanied companies on call outs during the day and the night because some photographs were taken on board appliances travelling through the streets. In addition, there are images of men loading hoses onto engines, hoisting and climbing ladders, holding hose nozzles, inspecting fire damage inside houses, and even of members of the public watching the men in action. Parks provided several individual portraits – such as driver Paul Honesty, who was a former grid star at Howard University, at the wheel of his engine – as well as group shots. The men appear at ease in Parks' presence. One particularly striking shot was taken at night from a high vantage point looking down on three men preparing to mount a ladder with a hose. The photograph is almost completely black except for the firemen's helmets and the line of a convoluted hosepipe with the result that image resembles an abstract painting. Another photograph, taken out of doors at night probably with a flash, shows two fireman gripping a rail while travelling on the back of a fire engine with coiled hoses at hand. When the truck approached a fire, one – 'the layout man' – jumped off and jerked the hose into the street in order to connect it to the nearest hydrant.

Over 40 of Parks' photographs of Firehouse Station No. 4, which were clearly intended to demonstrate what a positive contribution African-Americans were making to society, are preserved in the Prints and Photographs Division of the Library of Congress.

## 148 1944

### Norman Rockwell, Fireman

Rockwell (1894–1978) was the most famous and popular American illustrator of the twentieth century. His precisely detailed, illusionistic pictures with their heart-warming narrative scenes of American small-town life graced the covers and pages of such mass circulation magazines as *American Magazine*, *Look* and, most famously, *The Saturday Evening Post* (founded in 1728); consequently, they reached huge audiences.

Rockwell's interest in the theme of firefighters was first demonstrated in his painting *To the Rescue*, which appeared on the cover of *The Saturday Evening Post* for 28 March 1931. It depicts a side view of two figures and a dog dashing across the picture space from right to left towards a fire that is out of frame. (Hence, Rockwell breaks with the magazine cover convention that a frontal view should be presented.) The tallest figure, a fireman dressed in a red hat and a red shirt, has a large axe slung over his shoulder. Alongside him is a male youth carrying a red fire bucket.

Over a decade later, Rockwell's amusing painting *Fireman* (aka *Fire!* and *No Smoking*) was reproduced on the cover of *The Saturday Evening Post* for 27 May 1944. It was thus created during World War II but looks back to the late nineteenth century: the moustachioed fireman, a head-and-shoulders portrait set against a dark brown background, is mature in years and is wearing a old fashioned, red fire hat with the figure 1890 – a date? – and the number two on it. He also has on a red shirt (the volunteer companies of New York were fond of this colour) and is holding a speaking horn (presumably a sign of his rank as a foreman) against his chest. His right hand is thrust inside the fold of his shirt.

Actually, the painting is not simply a portrait but a painting of a portrait hanging on a wall covered with green, striped wallpaper. On a ledge below the painting, a cigar has been left to smoulder. A wisp of smoke rises vertically and causes the fireman to look down, flare his nostrils and grimace with displeasure at the carelessness which could so easily result in a blaze. One senses that here is an ancestor figure – a professional, experienced firefighter – who disapproves of the behaviour of later, affluent generations.

The portrait has a gold frame decorated with fire-fighting emblems: axes, hats, hoses, hydrants, ladders, lamps and horns, and the letters F.D. (Fire Department). In fact, Rockwell was inspired to paint this picture after discovering the frame in a junk shop in Troy, New York. He then rented the hat, red shirt and horn in New York. His model, however, was not a veteran fireman but the president of one of America's largest book publishers. As

148 Norman Rockwell, *Fireman*, (1944). Painting used as a cover illustration for *The Saturday Evening Post*, 27 May 1944. Printed by permission of the Norman Rockwell Family Agency © 1944 the Norman Rockwell Family Entities

was his habit, Rockwell took photographs of him and used them for reference purposes.

To supporters of avant-garde art, Rockwell was a commercial artist, an academic naturalist, a populist and a sentimental storyteller who did not merit a place in histories or museums of modern art. However, as this painting demonstrates, Rockwell was a more complex artist who was fully conscious of the mechanics of pictorial rhetoric. By being a picture-within-a-picture, *Fireman* calls attention to its status as a representation; furthermore, it is surreal because in response to the lighted cigar in the 'real' space outside the portrait, the fireman inside the depicted space of the portrait 'comes alive'. Today, prints of Rockwell's images can be purchased for displaying on walls; they also appear as decoration on neckties.

Another illustrator commissioned by *The Saturday Evening Post* who painted several detailed images featuring fire departments, fire stations and fire engines during the 1940s and 1950s was Stevan Dohanos (1907–94).

Towards the end of his life Rockwell returned to the theme of firefighting with an idealised scene entitled *The American LaFrance is Here!* It showed a small-town, wooden firehouse and the excited response of locals to the arrival of a powerful new fire engine. Rockwell lived in Stockbridge, Massachusetts and the vintage fire station he depicted was the town's Hose House No 1.

In March-April 2002, *The Saturday Evening Post* published a nostalgic article by Maynard Good Stoddard saluting those illustrators who had tackled the subject of firefighting in small-town and rural America, generally in a humorous or sentimental manner, in earlier issues.

## 149 — 1952

**Unknown Photographer, Richard Widmark and Jeffrey Hunter in a scene from Red Skies of Montana**

## 150

**Unknown illustrator, Dust jacket of Smoke Jumpers by Nels Jorgensen**

## 151 — 1957

**Nicolson, Front cover of The Wide World**

In 1952, the Hollywood studio Twentieth Century Fox released an action/drama/adventure film, entitled *Red Skies Of Montana* (aka *Smokejumpers*), about the work of American firefighters who tackle forest fires. Joseph M. Newman directed the Technicolor movie, Harry Kleiner wrote the script based on a story by Art Cohen, and its stars were Richard Widmark, Jeffrey Hunter and Richard Boone.

149 Unknown Photographer, Richard Widmark and Jeffrey Hunter in a scene from *Red Skies of Montana*, (1952). Publicity still © Twentieth Century Fox

Widmark (b. 1914), an actor who is more compelling playing giggling sadists than heroes, had the lead role as Cliff Mason, the foreman of a small crew of

150 Unknown illustrator, Dust jacket of *Smoke Jumpers* by Nels Jorgensen, (London & Glasgow: The Children's Press, n.d.). Author's collection

151 Nicolson? (Illustrator), Front cover of *The Wide World*, (London: George Newnes Ltd., Vol 118, February 1957). Author's collection

smokejumpers – men (and now women) who parachute into the forests to clear fire breaks or lines. Initially, he survives a wildfire but cannot remember how he managed to escape while all his men died. Hunter plays Ed Miller a rookie smokejumper who suspects Cliff of cowardice because his father Pop Miller was one of those who perished. Eventually, during a climatic conflagration, Ed physically attacks Cliff but the latter wins the fight and rescues all his men against the odds by getting them to shelter in foxholes he had ordered them to dig.

To some degree, the film resembles a documentary: there is a voiceover commentary at the beginning explaining how the US Forest Fire Service operates; there are scenes of training for parachute jumps at the base camp; we see the lookout watchtowers, aircraft, helicopters and radio equipment employed and we watch the smokejumpers clear brush, dig fire lines, fell trees and set backfires. There is also some information about the behaviour of forest fires. The final scenes in which the smokejumpers struggle to survive while surrounded by flames are the most authentic and gripping.

The plot of the film was loosely based on a real tragedy: the Mann Gulch Fire, Helena National Forest, Montana on 5 August 1949, in which 13 smokejumpers died. Some of the outdoor scenes were also filmed during 1951 at several locations near Missoula, Western Montana and footage of actual fires in California and New Mexico was incorporated. The film's technical advisor was Fred Brauer, an expert on smokejumping, who also liased for the Forest Service, which provided personnel and technical support for authenticity's sake. By no means a cinematic masterpiece, *Red Skies of Montana* is a competent B-movie that is both entertaining and educational. Naturally, the film remains popular with residents of Missoula.

In addition, during the 1950s, writers of adventure books for boys and of factual magazine articles for men were attracted by the subject of smokejumping and illustrators seized on the dramatic and dangerous moments when firefighters dressed in jumpsuits parachute from aircraft. Witness, for example, the dust jacket of Nels Jorgensen's book *Smoke Jumpers* (1954), which was published in the United States and Britain, and the cover of *The Wide World: The True Adventure Magazine for Men* (February, 1957), a British monthly. The latter image heralds an article entitled 'Smoke-jumpers of Silver City' by Starr Jenkins, a former smokejumper.

## 152 1955

## Arlen Kashkurevich, Portrait of a Fireman

In Russia, during the late nineteenth century, there was vogue for social realist painting and sculpture similar to that being created in Britain and France. Under the dictatorship of Stalin, in the 1930s, modern abstract art was condemned and socialist realist figuration based on nineteenth-century academic principles became the only style approved by the state. However, some pictures were enlivened by impressionist-type brushwork. Budding artists received a traditional education in state art schools and, if they showed promise, they were ensured commissions and employment through membership of the Union of Artists.

Arlen Kashkurevich, painter and book illustrator, was trained at the Belorussian Theatre and Art Academy in Minsk, joined the Union of Artists in 1960, and was made an Honoured Artist of the USSR in 1986. He has since been awarded the title 'People's Artist of Belarus'. (Belarus, previously a Soviet Republic, became independent of the USSR in 1991.)

One prized value or goal of socialist realism was '*narodnost*' (people-ness), the presence in art of the best qualities of humanity that could be enjoyed by the masses, and it is evident in Kashkurevich's 1955 portrait. Although the sitter is wearing a brass helmet and a protective overcoat, he seems to be posing in a studio rather than being on duty. Indeed, his mature age and the his-

152 Arlen Kashkurevich, *Portrait of a Fireman*, (1955). Painting, oil on canvas, 66 x 50 cms. Photo courtesy of Rodan Enterprises, Ilford, Essex

toric helmet suggest that he is a retired firefighter. His facial characteristics and moustache means that he resembles the man in Norman Rockwell's painting of 1944. (He also looks somewhat like Stalin.) The tough, experienced Russian firefighter has a resolute but rather melancholy expression on his face. His gaze does not meet the viewer's and he seems introspective as if reflecting on some harsh experiences. Kashkurevich eschews the craven flattery found in propaganda-type images of leaders and the exaggeration characteristic of Stakhanovite 'heroes of labour' representations; he also avoids the optimism and utopianism of so much Soviet art. He demonstrates that, even within the bounds of socialist realism, a sympathetic humanist portrayal of an actual individual was possible.

## 153 — 1962

### John Berry, British firemen attaching hosepipes to a hydrant, an illustration from The Fireman by Vera Southgate & J. Havenhand

During the 1960s, the British publisher Wills & Hepworth Ltd. issued a number of books for children entitled 'People at Work' as part of its best-selling Ladybird 'Easy-Reading' series. William Murray, an author and headmaster, had devised the latter to assist pre-school and primary school children to learn to read. Predictably, the Fire Service was one subject described in plain prose and seductive colour plates. The images favoured by the publisher were photographic in character and Berry, who illustrated *The Fireman*, regularly worked from photographs. During the 1930s, he trained as a fine artist at the Royal Academy Schools, was then employed in advertising and later became a society portrait painter.

According to Michael Bracewell, the Ladybird books contained 'romanticised images' depicting 'a postwar world of order and prosperity, at once idyllic and hierarchical – a place of safety'. Berry's style is certainly bland and his depictions of calm, efficient professionals must have been reassuring to both parents and their offspring. The images also depict a predominantly masculine profession. Only one female appears and she is a radio operator located in a fire station.

Rather surprisingly, in 2004 these images became

153 John Berry (illustrator), *British firemen attaching hosepipes to a hydrant*, an illustration from *People at Work: The Fireman* by Vera Southgate & J. Havenhand (Loughborough: Wills & Hepworth Ltd., 1962), 11.1 x 17.1 cm. Author's collection. © Ladybird Books Ltd. 1962

154 Charles Moore, *Firemen Trying to Dislodge Demonstrators, Birmingham, Alabama*, (1963). Photograph, 16.5 x 24 cms. © Charles Moore/Blackstar Agency, New York

fashionable as art: originals by Berry and other Ladybird illustrators were exhibited and offered for sale at the Simon Finch Art gallery in London. The organisers perceived a similarity between these images and photorealist paintings being produced by young contemporary artists.

## 154 — 1963

### Charles Moore, Firemen Trying to Dislodge Demonstrators, Birmingham, Alabama

This dramatic action photograph shows that not all the tasks firefighters are called upon to perform benefit the public. Charles Moore, a freelance photographer who worked for *Life* magazine during the 1960s and who specialised in documenting the Civil Rights movement, took it in Birmingham, Alabama on 3 May 1963 at a demonstration calling for the desegregation of city facilities. A team of white firemen had been ordered by Public Safety Commissioner Eugene 'Bull' Connor – an intransigent defender of racial segregation – to turn their hose – equipped with a high velocity 'monitor gun' nozzle – against black protestors, many of whom were teenagers. They huddle on the sidewalk trying to protect themselves from a powerful jet of water delivering 100 pounds pressure per square inch. The camera's position behind the hosepipe crew places the viewer on the side of the authorities committing an act of violence against peaceful protestors. Another photograph showed three young protesters trapped in a doorway by a jet of water.

In this instance, therefore, a fire department was used to impose the will of an oppressive local government authority. The next day Connor again ordered the firemen to use their hoses but to their credit they refused. One disgusted fireman later remarked: 'We're supposed to fight fires, not people.'

Moore (b.1931) was himself a white southerner from Alabama. He started taking photographs at the age of 14 and, after serving in the US Marines, he studied photography at the Brooks Institute in California. At first, his interest was in beauty and fashion photography but later he switched to photojournalism. Moore became a supporter of Civil Rights after witnessing the violence and ugliness of racism at first hand. Even he was threatened and physically attacked by racists. On 17 May, *Life* magazine devoted eleven pages to Moore's photographs of the pressure hose images and others showing police dogs attacking demonstrators. (Andy Warhol appropriated some of the latter for his silk-screen paintings.) The photographs shocked many Americans including politicians in Washington DC and provided evidence of the need for social change and in this way helped to bring it about.

## 155 1967

### Unknown photographer, Cyril Cusack (left) and Oskar Werner (right) playing firemen in the film Fahrenheit 451

In 1951, the American writer Ray Bradbury wrote a short story entitled 'The Fireman' for *Galaxy Science Fiction* magazine. Two years later, he expanded it into a novel, which was to become a bestseller and a school textbook. Joe Mugnaini illustrated the book and its odd title – *Fahrenheit* 451 – referred to the temperature at which paper burns. Universal Pictures made the novel into a 112–minute, Technicolor film in 1966–67. It was filmed in England at Pinewood Studios by British cinematographer Nicholas Roeg. Exterior shots show a clean but sterile environment of tower blocks, low-rise suburban housing and an overhead monorail. The distinguished French filmmaker François Truffaut directed it and co-wrote the script, and the Austrian actor Oskar Werner starred as fireman Guy Montag. Cyril Cusack gave a brilliant performance as Beatty, Montag's Captain, a man who alternates between charm and bullying. Both dress sinisterly in black uniforms and wear shiny, brown, beetle-like, close-fitting helmets. The Captain also carries a sidearm. Their fire engine and firehouse are painted a lurid red. The former is equipped with a hosepipe that delivers kerosene instead of water and the latter has a pole one can slide up as well as down.

155 Unknown photographer, Cyril Cusack (left) and Oskar Werner (right) playing firemen in the film *Fahrenheit* 451, (1967) Publicity/film Still. © Universal Pictures

In this film, the representation of a fireman was unusual because first, he is a fictional character living in the 24th century; and second, he does not put out fires, he starts them! According to the plot, a totalitarian government of a future society controls its zombie-like populace via drugs and continuous bland television. Writing and books are banned because they cause mental confusion and foster critical thought and therefore make for unhappy subversive citizens; consequently, Montag's job is to find and destroy all books. Fire brigades are not needed for traditional purposes because, supposedly, all new buildings are fireproof.

On one raid, Montag is disturbed when an old woman with a large collection of books burns herself to death rather than live without them. Despite the promise of promotion, he becomes dissatisfied with his alienated existence and curiosity prompts him to read some books he has secretly confiscated. After his conformist wife (played by the British star Julie Christie) denounces him, Montag resists arrest by killing Captain Beatty with a flamethrower. He then joins other social outcasts – 'The Book People' – who live in the countryside. Even the latter destroy books (as physical objects) while 'preserving' them by memorising and reciting them.

Although *Fahrenheit* 451 has its admirers and is thought provoking compared to mindless Hollywood blockbusters, it is a rather slow and melancholy film. Its pace and style emulates a Hitchcock thriller but it lacks the latter's tension and excitement. The image it presents of the fireman, of course, is a world-turned-upside-down one.

## 156 1973

### Unknown photographer & graphic designer, Actor Richard Roundtree on the front cover of the video version of Firehouse

Although television transmissions occurred before World War II, television only became a mass entertainment medium in Britain and the United States during the 1950s. Richard C. Yokley, the author of a history of television programmes about firefighters, has claimed that the first example he can trace was *Alarm* produced in

156 Unknown photographer & graphic designer, Actor Richard Roundtree on the front cover of the video version of *Firehouse*, (1973). Telemovie. Production companies: Stonehenge and Metromedia for ABC Television. Design of video packaging © 1999 Gemstone Entertainment

Los Angeles in 1954. Since then, of course, television production companies have generated dozens of documentaries and fictional programmes and series about the subject. Only a few examples can be considered here.

Richard Roundtree (b. 1942), an African-American actor, became a star playing the role of the tough, ultra-cool private eye John Shaft in the 1971 movie *Shaft*. Two years later, he starred as rookie fireman Shelly Forsythe in a film made as a pilot for a television series entitled *Firehouse* (aka *Night Watch*, 1987), which is set in an American firehouse at a time of social change and unrest. (The firehouse was supposedly located in New York but the film was made in Los Angeles using an ex-firehouse dating from 1910 located at 225 East 5th St.) Alex March, a veteran of many television series, directed the film and its script was written by Frank Cucci. Andrew Duggan played Jim Parr, a sympathetic Captain, while Vince Edwards played Spike Ryerson, a brave, experienced but racially prejudiced fireman.

When the film begins, an all-white male fire company – Engine Company No. 23 – loses Doyle, one of its best men while fighting a blaze. After Doyle's funereal, his mourning and bitter colleagues express hatred towards African-Americans because they blame black arsonists for so many of the fires and malicious false alarms they are having to attend. Much to their disgust, their dead friend's replacement is Forsythe who is suspect in their eyes because they think he will be loyal to his race rather than to the fellowship of the firehouse. Forsythe, who fears he is a token of desegregation and affirmative action policies, does not help matters by behaving in a truculent manner. He is unwilling to cook meals and when told the fire company is as close as 'a family', Forsythe says he already has one (he is married). Predictably, Forsythe has to endure tedious jobs, hazing – mostly from Ryerson – and a fistfight before his bravery and professional performance as a firefighter wins their respect. In one memorable scene, Ryerson rescues an unconscious Forsythe from a burning building and then gives him mouth-to-mouth resuscitation. Eventually, it is Forsythe who catches the principal black arsonist thereby demonstrating that he is a first and foremost a firefighter.

This pilot film was first screened on ABC television on 2 January 1973 and later issued on video (Gemstone Entertainment) and DVD (Passion Productions). Shot with gritty realism, it holds the attention and is competently acted; the fire-fighting scenes appear authentic. (Credits are given to the fire services of Boston, Cleveland, Los Angeles and New York City, and to technical advisor Captain Scott Smith of the LA Fire Department. Smith also played the part of a battalion chief and supplied a fire engine.) Furthermore, the culture, routines and tensions of firehouse life are convincingly portrayed, although the reality of the humiliations and isolation actual African-American firefighters endured for many years is not fully conveyed and overall the plot seems thin.

In 1974, a 13–episode/30–minute long, television series entitled *Firehouse* located in a Los Angeles firehouse was transmitted but neither Roundtree nor Edwards participated.

*Firehouse* was clearly a cultural response to topical events. During the late 1960s and 1970s – a period known to New York firefighters as 'the war years' – fires and false alarms in inner city black ghettos increased dramatically due to crime, greed, poverty, urban decay, riots and terrorist bombs. White property owners torched their buildings to claim insurance monies and angry blacks set fires as a form of protest. When the hard-pressed fire companies answered alarms, they were sometimes attacked with bricks and bullets. At this time, American fire departments were encouraging ethnic minority recruits to make the workforce more representative of the wider society.

From March to May 1999, Roundtree again played the part of a fireman-rescuer in the eight-episode, television series *Rescue–77* (Spelling Television/WB network). As Captain Durfee, he commands a small rescue and paramedical team assigned to Station 77 in Los Angeles. The series writer and producer was Gregory Widen, a retired firefighter-medic.

## 157 — 1974

**Unknown photographer, Steve McQueen plays a Battalion Chief in Irwin Allen's The Towering Inferno**

## 158 — 1974

**The Tower aflame. Publicity still**

Hollywood studios have made several fictional feature films about fires and firefighting. *The Towering Inferno*, one of the best, was produced by two studios – Twentieth Century Fox and Warner Bros – and was based on two similar novels – *The Tower* (1973) by Richard Martin Stern and *The Glass Inferno* (1974) by Thomas N. Scortia and Frank M. Robinson – to which the studios had acquired rights. In addition, a real event may have influenced the plot: on 1 February 1974, the Joelma Building, a new 25–storey office block in Sao Paulo, Brazil caught fire. Hundreds of occupants were trapped on the upper floors and roof. Rescuers used a helicopter and lines fired by a harpoon gun from neighbouring buildings to save some victims. (These tactics also appear in *The Towering Inferno*.) However, despite the determined efforts of the local fire brigades, 227 people died in this fire, which was aggravated by inflammable plastic building materials and an inadequate escape staircase.

Irwin Allen, so-called 'King of the Disaster Movies', produced the movie and directed action sequences while John Guillerman directed drama sequences; Stirling Silliphant wrote the script. The cast was crammed with famous actors: Fred Astaire, Richard Chamberlain, Faye Dunaway, William Holden, Jennifer Jones, Steve McQueen, Paul Newman, Robert Vaughn, Robert Wagner and even O.J. Simpson.

The brand new glass skyscraper – dubbed 'the tallest building in the world' – is supposedly situated in San Francisco and so exteriors and documentary-style footage of fire engines were shot there. Filming also took place inside the lobby of the City's Hyatt Regency building. However, most of the movie was filmed on sets constructed at Fox's Century City studios in Los Angeles.

This well-acted, Oscar-winning movie concerns the fate of a motley collection of characters trapped at the top of a 138–floor skyscraper by a fire, which breaks out on the 85th floor. It is caused by short-circuits in inferior wiring due to skimping on the building costs. Predictably, the contractor responsible for ignoring the architect's electrical specifications, played by Chamberlain, dies a horrible death. Fire alarms alert the firehouses and we see men sliding down poles, rigs rushing through the streets, hoses being deployed, and the Battalion Fire Chief Michael O'Hallorhan – played by McQueen – setting up his command post and consulting plans with the architect Doug Roberts – played by Newman. Many of the subsequent firefighting scenes appear authentic and, given the range of techniques shown, the movie could be used for training purposes. The Fire Chief, a model of cool efficiency and matter of fact bravery, also cares for the men he has to direct into danger. Only later, do his feats border on the superhuman and seem far-fetched.

The film is franker about the effects of fire than some documentaries: civilians and firefighters are injured and die, some become fireballs that crash though windows and fall from the building. A scenic elevator made from steel and glass located on the outside of the skyscraper becomes the focus of some dramatic scenes as the Fire

157 Unknown photographer, Steve McQueen plays a Battalion Chief in Irwin Allen's *The Towering Inferno*, (1974). Publicity still. Copyright © 1974 Twentieth Century Fox & Warner Bros

158 *The Tower aflame*. Publicity still for Irwin Allen's *The Towering Inferno*. Copyright © 1974 Twentieth Century Fox & Warner Bros

Chief dangles from a helicopter to perform an improbable rescue. In a desperate last act, he and the architect blow up the building's water tanks. The ensuing flood drowns the fire and the structure survives.

Battalion chiefs from the San Francisco and Los Angeles Fire Departments acted as technical advisors on the film and professional firefighters stood by when sets were engulfed in flames. During fire scenes, stuntmen wore innovative, lightweight, fire-retardant suits underneath everyday clothing.

*The Towering Inferno* began with a dedication to the heroism of firefighters and was promoted as a tribute to them. It quickly recouped its $14 million production cost by grossing $116 million on release. Worldwide distribution, video and DVD editions, and television screenings ensured that millions viewed the film. Some critics found it overlong (165 minutes) and disliked it, but it proved popular with the public and is still rated as one of the best disaster movies of the 1970s.

In the light of what happened to the Twin Towers on 9/11, *The Towering Inferno* was a highly prescient movie because it depicted the great difficulty of defeating fires that occur high up in skyscrapers. (One wonders if it inspired the terrorists.) Early in the movie, the Fire Chief reminds the architect that fire departments cannot effectively fight fires in skyscrapers beyond the seventh floor. At the end of the film, after passing several corpses of his men, the Chief specifically raises the question of design with the architect. He remarks: 'One of these days we are going to kill 10,000 in one of these firetraps and I'm going to be eating smoke and bringing out bodies until someone asks us how to build them.' Unfortunately, many city authorities, property developers and architects have since ignored this advice – they still want and build very tall buildings in spite of all the dangers they pose.

Within the fire service, there are many reports about high-rise fires and discussions of the best methods of tackling them. For instance, Vincent Dunn a retired Manhattan Deputy Chief, has issued a 25–minute video entitled *High Rise Firefighting* giving advice on strategies for battling such fires and evacuating people from upper floors.

## 159 — 1977

**Peter Johns, London Firefighters on Picket Duty**

## 160 — 1977

**Peter Kennard & Peter Harrop, Front cover of the newspaper Socialist Challenge**

## 161 — 1987

**Richard Manning, United Firefighters**

Most visual representations of firefighters do not show them as trade unionists or as militant demonstrators. Such images do exist but they are at odds with the familiar heroic ones. In Australia a fire brigades employees

159 Peter Johns, *London Firefighters on Picket Duty*, (1977). Photograph. © Guardian Syndication

# Socialist Challenge

No. 25 8 DECEMBER 1977 15p

SUPPORT TH[E] F[IREMEN]

Photomontage: PETER KENNARD

Photo: PETER HARRAP (Report)

TUC
No Support for the Firemen
Len Murray (Gen Sec)

## Len Murray: CALLAGHAN'S GREEN GODDESS

## YOUR PAPER DOES IT AGAIN

Last week Socialist Challenge published exclusive revelations about Willesden Magistrate Dorothy Oakley. We accused Oakley of bias against Grunwick pickets.

Within 24 hours of publication, Oakley was hearing no more Grunwick cases.

see page 9

## You can help the Firefighters by

*Building meetings of shop stewards from every workplace, through trades councils and stewards' committees, to plan local action.
*Building for local solidarity strike action to attend oemonstrations with all other workers in support of the firefighters.
*Winning your workplace to 'adopting' a local firestation, to make sure that the firefighters strike remains solid and merry over Xmas.
*Sending resolutions to your union executive and the TUC demanding massive financial donations to aid the strike, and a campaign in their support.
*Lobby the TUC General Council on 21 December to demand they support the strikers and campaign against the 10 per cent limit. Battersea and Wandsworth Trades Council Fireman's Support Committee have already called for a lobby — see your union branch or stewards committee does too.

160 Peter Kennard (montage) & Peter Harrop (photograph), Front cover of the newspaper *Socialist Challenge* (No. 25, 8 December 1977). Reproduced courtesy of Peter Kennard

161 Richard Manning, *United Firefighters*, (1987). Painted banner, acrylic on canvas, 2m 44cms x 3m 5cms. Trade union banner of the United Firefighters Union – Victoria, housed in the branch office in Fitzroy, Victoria, Australia. Photo courtesy of Peter Marshall and Matt Morgan of the UFU

union was formed as early as 1911 and the current United Firefighters Union of Australia, Victoria branch commissioned, in 1987, a colourful and dramatic banner. It was painted by artist and firefighter Richard Manning, who was clearly inspired by Eugène Delacroix's famous revolutionary canvas *Liberty Leading the People* (1830). In Britain, a national Firemen's Trade Union, later called Fire Brigades Union (FBU) was formed in 1918. Senior officers belong to a professional association – the National Association of Fire Officers – that eschews strike action.

Since firefighters provide a life-saving service to the public, they are extremely reluctant to take industrial action and rarely do so. Nevertheless, at times, rank-and-file British ones have felt such action was justified if only to preserve that service and to obtain fair pay and conditions. In a history of the FBU entitled *Forged in Fire*, Victor Bailey wrote:

> The first official strike in the Fire Brigades Union's 59–year history began on 14 November 1977, and ended, after nine weeks, on 16 January 1978. If one image encapsulates the strike, it must surely be that of firemen on picket duty, huddled around braziers outside closed fire stations, acknowledging the signs of public support in the form of financial donations and tooting car horns.

In 1977, the FBU sought a wage comparable to that of skilled industrial workers plus 10 per cent to take account of the skills qualified firefighters possessed and the dangers they faced. Despite considerable public sympathy, they faced an obdurate Labour Government unwilling to breach an incomes policy it had imposed on the nation's workers.

Peter Kennard (b.1949, London), a left-wing British artist who specialised in photomontage during the 1970s, supported the striking firefighters with a critical montage showing three at a public meeting (photo by Peter Harrop) juxtaposed against an image of Len Murray, General Secretary of the Trades Union Congress, who was refusing to help the strikers. The montage was published on the front cover of the newspaper *Socialist Challenge* (No. 25, 8 December 1977), organ of the International Marxist Group.

After a long and bitter struggle in which the firefighters evinced great determination and solidarity, they conceded defeat and returned to work following a modest wage increase. However, they did achieve a pay formula that was to last for over two decades.

As Bailey noted, the image of firefighters standing around a lighted brazier quickly became a pictorial cliché. It is a rather ironic image because firefighters normally extinguish flames rather than cause them. Journalist Catherine Bennett, writing in 2002 about the publicity value of the brazier, described it as 'an animated focal point rich in mythic and historic associations which also acts as a permanent reminder of the fires they [firefighters] might be out extinguishing were it not for the frigid meanness of the employers.' Most fire stations are scattered across urban areas and so strikers outside them with their 'string of inner-city beacons' are highly visible; consequently, the public soon becomes aware of any strike and the strikers in turn soon become aware of the public's support or lack of it.

## 162 1977

## Jill Freedman, Injured Fireman

In 1977, Doubleday & Company Inc., of Garden City, New York published a book of brownish photographs by Jill Freedman entitled *Firehouse*, which was produced as a tribute to all firefighters. Two colour photographs – portraits – appeared on the book jacket. The volume included a personal note by Freedman plus a brief essay written by the noted author Dennis Smith who, at the time, was a full-time firefighter assigned to Ladder Company 61 in the Bronx. Smith describes his work during one daytime shift and so his text complements Freedman's images rather than directly commenting on them.

Freedman (b. 1939) is now an experienced and respected freelance photojournalist and social documentarist whose images have been published in major magazines; they have also been displayed in art galleries and museums. In her youth, she studied sociology and cultural anthropology but during the 1960s, she taught herself photography and then documented poor people's campaigns. She is noted for a series of photo-essays representing particular communities and their everyday lives. Her intimate portrayals are achieved by spending a long time within communities so that mutual learning can take place. As a child, Freedman had wanted to be a firefighter and later she photographed a firefighter rescuing a cat. She had always regarded firefighters as gentle heroes performing a valuable public service and so she decided to record with her camera the daily lives of a company based in the South Bronx. Like David Parks' earlier photo-essay, Freedman generated a wide range of images: individual and group portraits, fireman performing chores and relaxing inside the firehouse, fighting fires, rescuing and giving medical attention to victims, talking to children and other members of the community in the street. Informality is a quality of many of her compositions.

As the photograph reproduced demonstrates, she also documented injured firefighters and moved in close to do so. She also photographed a firefighter's funeral. Such images are a salutary reminder of the risks they take and contrast with the majority of representations of firefighters, which depict them in control saving others.

162 Jill Freedman (photographer), *Injured Fireman*, (1977). Photo courtesy of Jill Freedman

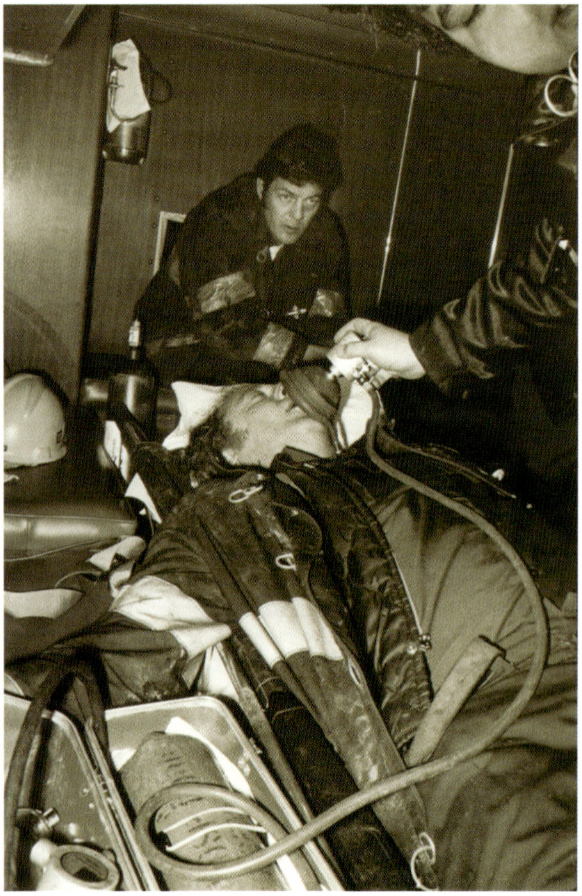

163 Press Association photographer, *A Lift for Pascale ...*, (1979). Publicity photograph, London. Photo courtesy of the Press Association

| 163 | 1979 |
|---|---|
| **Press Association photographer, A Lift for Pascale** | |
| **164** | **1935** |
| **Unknown illustrator, The Fireman's Lift** | |

The photograph of a so-called 'fireman's lift', dating from October 1979, shows two men from the London Salvage Corps (LSC) holding Pascale Paintendre, a French airline reservations agent, between them by linking arms to form a cradle for her to sit on. (Insurance companies had combined to establish the LSC in 1866. Its men attended major fires to protect property from excessive water damage and looting. The LSC was disbanded in 1984.) On the left is Terry Wilds wearing LSC clothing and on the right is Frank Gardener dressed as an American firefighter. This staged photograph was taken by a Press Association photographer to publicise an exhibition held at the Royal Exchange, London as part of

164 Unknown illustrator, *The Fireman's Lift*, (1935?). Detail of illustration in *The Picture Book of Firemen* (London & Melbourne: Ward, Lock & Co., Ltd)

National Fire Safety Week. Educating the public regarding fire safety measures is one of the essential but mundane tasks fire officers have to perform. Ms Paintendre was clearly intended to add glamour to the picture so that newspapers would print it.

Because of the heavy gear and loads firefighters have to carry, firefighting is a physically demanding occupation and so applicants need to be fit and must be able to pass certain strength and stamina tests, one of which is the ability to lift and carry or drag an unconscious body. This is the origin of the term 'fireman's lift', (called 'carry' in the United States), which traditionally involved balancing a facedown body across the shoulders with one arm between or around the victim's legs. (Captain Shaw, in his book *Fire Protection – A Complete Manual* ... [1876], described the fireman's lift in precise detail.) When women wanted to become firefighters, many men thought they would not be able to pass this test but, in fact, determined fit and strong women were able to do so.

## 165 — 1979

### Levon Konstantinovich Lazarev, Monument to the Firemen who Perished in the Great Patriotic War

## 166 — 1942

### Boris Artzybasheff, Fireman Shostakovich

During the USSR's Great Patriotic War against Nazi Germany, St Petersburg, then called Leningrad, suffered a 900–day siege (from September 1941 to January 1944) and was subjected to bombing, shelling and restriction of supplies. Starvation even forced some citizens to resort to cannibalism. Among over 600,000 casualties were the city's firefighters. Histories of the siege record that a shortage of fuel to heat buildings compelled the population to burn whatever they could find in makeshift stoves. This resulted in a large number of accidental fires. The fire service was frequently helpless because of frozen water pipes, a lack of petrol for appliances and because so many men were off sick.

In 1979, Levon Konstantinovich Lazarev, one of St

165 Levon Konstantinovich Lazarev, *Monument to the Firemen who Perished in the Great Patriotic War*, (1979). Monument/public sculpture, bronze. St Petersburg: Bolshoi Prospect, Vasilievsky Island, in front of Fire Station No. 9. Photo: Peter Y. Sobolev

166 Boris Artzybasheff, *Fireman Shostakovich*, (1942). Illustration, front cover of *Time* magazine, Vol. XL, No. 3, 2o July 1942. Reproduced courtesy of Time Life Pictures/Getty Images

Petersburg's leading artists (a sculptor, painter and graphic artist), was commissioned to create a monument to commemorate those who died on duty. He was born in Tbilisi in 1928 and graduated from the Leningrad High School of Art & Industry in 1955. Two years later, he became a member of the Russian Artists' Association and since then has exhibited widely both inside and outside Russia, and executed many public monuments in St Petersburg.

Lazarev's sculpture is unusual because of its dynamism and departure from the norms of socialist realism. Two helmeted firemen swathed in thick protective clothing are represented in action. Their faces are heroic but quite generalised. One fireman has his arms outstretched in a gesture of helplessness while behind him the second grasps a hosepipe in his right hand and an axe in his left. In a gesture of defiance, the axe is raised high in the air. Both figures balance precariously on an irregularly shaped support resembling an anvil. They appear to be falling along with a ladder and masonry. (However, at first sight one might think they were dancing!) Appropriately, the Monument has been situated in front of Fire Station No. 9, at the Bolshoi Prospect since 1995. Before that, it had been located inside the Station's courtyard.

One of Leningrad's fire wardens during the siege was the famous Russian composer Dmitri Shostakovich (1906–75). He guarded the roof of the Conservatoire of Music against incendiaries. His *Seventh Symphony* (the Leningrad) was composed during this period and was later performed in Britain and the United States. A photograph of Shostakovich dressed as a fireman was widely circulated and used to symbolise the heroic resistance of the people of Leningrad. Even the American magazine *Time* honoured him, in July 1942, with a cover portrait set against the burning ruins of Leningrad. The composer's head is represented in profile and his face has a severe expression. He wears spectacles and a fireman's metal helmet; the caption reads: 'Fireman Shostakovich: Amid bombs bursting in Leningrad he heard the chords of victory.' Some musical chords appear in the top right-hand corner.

Boris Artzybasheff (1899–1965), the artist responsible for this hyperrealist portrait, was himself of Russian origin. After fighting in the White Russian army, he moved to the United States in 1919 and worked for a New York engraving shop before becoming a freelance book and magazine illustrator. He created over 200 covers for *Time* between the mid 1930s and the 1960s. Among other Russians he depicted for the magazine's cover were Joseph Stalin and Nikita Khrushchev.

## 167 — 1990s

## Stefan Sträubig, Berlin firefighters standing next to the Feuerwehrehrenmal

## 168 — 1981

## Kurt Mühlenhaupt, Feuerwehrbrunnen

Mariannenplatz is situated in the Berlin district of Kreuzberg, which has long been a bohemian quarter where many artists live and maintain studios. In the square, there is a small park and a wall almost as tall as a human being with a carved surface simulating flames. The *Feuerwehrehrenmal* is a memorial site for Berlin's fire brigades, which hold wreath-laying ceremonies there to honour fallen comrades. Unfortunately, the memorial is often disfigured by sprayed graffiti. Today's Berlin Fire and Rescue Service had it origins in the Prussian Fire Brigade founded in 1851. During the 1990s, around 4000 professional firefighters served a population of 3.5 million. In addition, there was a volunteer force numbering 1300.

Just across a road from the *Feuerwehrehrenmal* are three, life-size, bronze firefighters by Kurt Mühlenhaupt, a German artist who has had a long career. He was born in

167 Stefan Sträubig (photographer), *Berlin firefighters standing next to the Feuerwehrehrenmal* (c. 1990s). Berlin: Mariannenplatz

168 Kurt Mühlenhaupt, *Feuerwehrbrunnen*, (1981). Fountain/public sculpture, bronze. Berlin: Mariannenplatz. Photos courtesy of Stefan Sträubig, Feuerwehrmuseum, Berlin. Photos courtesy of Martin Klein

1921 and from 1946 to 1948 he studied at Berlin's School of Fine Art. He is best known as a figurative painter and printmaker depicting ordinary people and urban scenes executed in a naïve style. He favours dark colours and so his pictures are rather gloomy and melancholy. His 1981 sculptures of firemen are therefore exceptional in his *oeuvre*.

The bronzes are stocky men dressed in late nineteenth century uniforms and helmets (Berliner Kappes), and they grapple with two writhing hosepipes attached to a fire hydrant near the edge of a low basin. In fact, they are parts of a fountain; appropriately, the hosepipes deliver the water. What is most striking about Mühlenhaupt's figures are their non-naturalistic faces: huge noses and protruding chins and lips. It seems this is one of the artist's trademarks. Not surprisingly, Berlin's firefighters were displeased to be represented in this caricatured fashion but it does serve to make the sculptures amusing, vivid and memorable.

## 169 — 1983

**'Hal' (Pat Hallahan), 'Looks like the electricity is still on ...' Cartoon, pen-and-ink line drawing, published in "When the Bells Go Down"**

## 170 — 2002

**Chris Brown, Fester the Fireman**

Some of the cartoonists who have made fun of firefighters have themselves belonged to the service. A British example is 'Hal' – Pat Hallahan – whose cartoons appeared regularly in *London Fireman*, the in-house magazine of the London Fire Brigade. Given the magazine's readership, Hal's jokes were primarily intended for internal consumption. For 16 years, Hallahan had been an LFB fireman and during the 1980s was based at station L23, Hainault in Redbridge. In 1983, he self-published a collection of his cartoons entitled "*When the Bells Go Down*", which is British firefighter parlance for bells sounding an alarm. Since Hal had first-hand inside knowledge, he was able to mock with self-depreciating and black humour many aspects of a firefighter's life that an outsider might have missed: jokes at the expense of the Union, officers' absurd instructions and the daily tasks and duties of firefighting. Several cartoons – such as the one reproduced – made light of the dangers of the

169 'Hal' (Pat Hallahan), '*Looks like the electricity is still on ...*' Cartoon, pen-and-ink line drawing, published in "*When the Bells Go Down*" (Ilford, Essex: Pat Hallahan, 1983)

170 Chris Brown, *Fester the Fireman*, (2002). Caricature, pen-and-ink line drawing and watercolour. Photo courtesy of the artist

job but, in addition, there were cliché-ridden ones about mothers-in-law and alluring females. All but one of Hal's firefighters are male and his main generic character is, in Gordon Honeycombe's words, 'the bug-eyed, spluttering, baggy, bumbling and much abused fireman'; consequently, this character contrasts sharply with the noble hero favoured by most visual artists.

Another unflattering, non-heroic, cartoon fireman-character created by a serving British firefighter is 'Fester the Fireman', a squat, smelly, dwarf-like individual with a huge nose and shifty eyes devised in the 1990s by Chris Brown based at Ringinglow Fire Station in South Yorkshire. Brown is a talented caricaturist with a surreal sense of humour. Fester, he claims, was born in 1952 following the marriage of Lt. Col. Sir Reginald Bulgingcrotch, a rich military buffoon, and Ethel Cretin, a scullery maid. Brown's images and comic strips of Fester have enlivened Christmas cards, pottery and a website entitled 'hedgepig'.

Harold 'Poony' Pickering is one American cartoonist who has satirised the fire service in a book entitled *Poonytoons: The 'Lighter Side' of the Fire Service* (1989). He drew cartoons about firemen's voyeurism, lack of culinary skills, mistakes with equipment, absurd bureaucracy and excessive workloads. Siding with rank-and-file firefighters, he also created images expressing their resentment at officious fire chiefs carrying clipboards and walkie-talkie radios.

## 171 — 1984

## Edd Hayes, Red Adair – The Hellfighter

## 172 — 1969

## Unknown photographer, John Wayne Katherine Ross and a firefighting scene from the film Hellfighters

Red Adair (1915–2004) was an exceptional American firefighter for several reasons: first, he was an individual whose fame extended far beyond the firefighting community; second, he was a specialist in extinguishing oil well fires, capping wild wells, and dealing with offshore blowouts; third, he operated in the private not the public sector; four, he established his own companies – the Red Adair Company Inc., the Red Adair Service & Marine Company and Adair Enterprises, Inc., – and so became a successful and wealthy entrepreneur; five, he invented advanced firefighting equipment and developed new techniques for the control of oil well fires; six, he was an advisor on a major Hollywood feature film.

Paul Neal Adair, nicknamed 'Red' because of his red hair, was born into a poor working-class family living in Houston near the oil fields of Texas. As a boy, he was a fearless street fighter and accomplished athlete. After school, he worked for the Southern Pacific Railway, and in 1938 for the Otis Pressure Control Company. During World War II, he served in a bomb disposal squad in Japan. Once demobilised, he learned how to control oil-well fires and blowouts from the pioneer Myron Kinley (1896–1978). Then, in 1959, he founded his first company, which undertook work internationally and completed highly dangerous tasks in deserts, jungles and on

171 Edd Hayes, *Red Adair – The Hellfighter*, (1984). Sculpture, bronze and marble, life-size portrait bust. Collection Red Adair Family. Photo courtesy of Edd Hayes, Texas

172 Unknown photographer, John Wayne, Katherine Ross and a firefighting scene from the film *Hellfighters*, (1968–69). Lobby card. Photo Copyright 1969 © Universal Pictures

North Sea oilrigs. In 1962, he gained respect within the industry by putting out a natural gas fire in Algeria, dubbed 'the Devil's Cigarette Lighter', which had burned for six months. Perhaps his most notable achievement occurred in 1991 when he and his team extinguished 117 burning oil wells in Kuwait, which had been set alight by retreating Iraqi troops during the Gulf War.

Adair's fame was spread by news reports, press profiles, television appearances, and by an authorised biography: *Red Adair: An American Hero*, (1989) written by Philip Singerman. In addition, a biographical documentary entitled *The Fireman: The Story of Red Adair* was released on video in North America in 1998. Adair's most recent company also maintains an Internet website about him: www.redadair.com During his lifetime, Adair received many awards, honours, medals and letters of recognition (some from American presidents) for his 'courage, perseverance and skill'. One citation declared: 'In an age said to be without heroes, you are an authentic hero.'

In 1984, Adair's daughter Robyn commissioned Edd Hayes to produce a bronze portrait bust of her father for his 70th birthday. Hayes was born in the 1940s and grew up on farms hero-worshipping cowboys. Then, during the 1960s and 1970s, he served in the U.S. Navy and worked as a rodeo rider and steer wrestler. Inspired by the Western or Cowboy art he saw in the Gilcrease Museum, Tulsa, Oklahoma, Hayes took up art and taught himself to draw and paint. His first sculptures date from 1981. He is now a successful Texan sculptor noted for his 'action' portraits of humans and animals, and rodeo monuments, executed in a naturalistic manner.

For his clay model of Adair, Hayes worked from photographs because Robyn wanted her gift to be a surprise and, in any case, his subject was in Spain at the time. The news photograph that Hayes selected captured a dangerous moment while Adair was fighting an oil-well fire. He was holding dynamite caps and shouting at a bulldozer driver to stop before it cut wires that would have caused an explosion; hence, the animated face, open mouth and the urgent pointing gesture of his left arm and hand. Although the portrait is dynamic and seems a good likeness, the inevitable lack of a background in a freestanding sculpture means that the context of black oil and flames erupting into the sky from a desert well is missing (compare illus 182).

The Hollywood movie that Adair advised on technical matters was entitled *Hellfighters* (Universal Pictures, released December 1968, directed by Andrew V. McLaglen) and starred John Wayne as Chance Buckman a reckless oilfield firefighter. The two men, one a genuine tough guy and one who played tough guys on screen, became friends during the shooting of the film in Houston and Wyoming in 1967. Although many believe the screenplay was based on Adair's life and personality, his biographer Singerman denies it. However, the spectacular action sequences showing Buckman and his team tackling oil-well fires, which were far more gripping and informative than the family feuds and love lives of the main characters, resembled some of Adair's exploits and the red overalls worn by the principal firefighters inevitably evoked Adair's nickname.

## 173 — 1986

## Uncredited photographer, Katherine Rogers as Josie in London's Burning

In Britain, in December 1986, London Weekend Television screened a lengthy action drama – *London's Burning* – written by Jack Rosenthal (1931–2004), directed by Les Blair and produced by Paul Knight, which was set in a fire station. Rosenthal researched the daily routine of firefighting by going on 'shouts' with crews from fire stations in Hornsey and Soho, London. The drama, which was later released on video, featured a variegated cast of characters and was well crafted. Its firefighting scenes were reasonably convincing. Even though very few firewomen were employed at the time, Rosenthal was keen to include one and so invented the character of Josie Ingham played by Katherine Rogers, an actress who took a rigorous training course in preparation for her role. Her arrival at the station causes consternation to the all-male watch but she is eventually accepted. Working at the station is one black firefighter, nicknamed 'Ethnic', who has to endure racial banter. At the climax of story, Ethnic is killed during an urban riot while trying to protect a colleague from a mob.

Two years later, a series began – produced by Paul Knight with scripts by various writers – which by 2002 had reached number 14. It follows the daily lives, on and off duty, of a crew called 'Blue Watch' located at 'Blackwall Fire Station', London.

The London Fire Brigade was persuaded to provide technical advice for the series' pyrotechnics and stunts, and in turn benefited in terms of positive publicity and soaring recruitment. The LFB was also able to insert fire safety messages. Commercial Union, an insurance company, has sponsored the series.

*London's Burning* has proved popular (audiences have reached as high as 18.9 million) and has generated spin-offs such as books, a board game, videos and websites maintained by fans. However, for viewers who are primarily interested in the work firefighters perform and the methods they use, the series is unsatisfactory because it devotes too much time to the personalities, habits and love affairs of its leading characters. Answering alarms – 'turnouts' – do occur of course and serve to enliven with action and danger the soap-opera banality of the rest of the programmes.

At least two other countries – Australia and the United States – have made similar drama series. *Fire*, for instance, transmitted during 1995 and 1996 by Australia's Network Seven, was a 50–minute, 26–episode series produced by Beyond Productions & Liberty Films. Written by Tony Cavanaugh and directed by Geoff Bennett, it was set in a Brisbane fire station and featured 'Morgan', the first female firefighter in the Queensland Fire Service.

In February 2004, a new, 20–part firefighting drama series entitled *Fireflies* (A Southern Star John Edwards Production for ABC TV and Austar) began transmission

173 Uncredited photographer, *Katherine Rogers as Josie in London's Burning*, (1986?). Photo Copyright © London Weekend Television

in Australia. Described as 'compelling' and 'realistic' but also 'trite', the series depicts the lives of members of a rural fire brigade in Lost River (a small town) as they fight bushfires and each other during a torrid summer. Volunteer organisations such as the Rural Fire Service (RFS), which has 250,000 members in Australia, help to bind local communities together. Created and written by John O'Brien, *Fireflies* was produced by John Edwards and directed by David Caesar and Geoff Bennett. Real life firies played the part of extras in the drama and thought its fire scenes looked authentic.

*L.A. Firefighters*, a six-episode series set in a Los Angeles' firehouse, was screened by the Fox Network in 1996. Created by executive producer Gordon Griesman and directed by Cliff Bole and others, it starred Jarrod Emick as Captain Jack Malloy. Christine Elise played a female firefighter who is the daughter of a Battalion Chief who disapproves of her choice of profession. Although the series prompted criticism and laughter from the professionals and was dubbed '*Baywatch* in bunkers' and judged 'one of television's worst shows', a fan has commemorated it with a website. Twelve new episodes were made with the title *Fire Company* 132 but these were only shown outside the United States.

## 174 — 1987

**Rob Lee, John Walker & Ian Frampton**
**Fireman Sam: The Hero Next Door**

## 175 — 1959

**Virginia Hunter & Marge Opitz,**
**Fireman Joe**

As explained in the Introduction, there are many tales about, and images of, firefighters targeted at children both before they attend primary school and after. Invariably, the tales/images involve the creation of a 'character' or cartoon figure, in effect a commercial brand, with names such as 'Fireman Joe', 'Fireman Small' and 'Freddy the Fireman'. As we have seen in the case of Big Mose, the New York fire laddie, some children's characters are based on real people but the majority are invented.

In Britain, one of the most popular of such characters has been Fireman Sam, who was conceived in 1985 by Dave Gingell and Dave Jones, two ex-firemen from Kent. With the help of animator Mike Young, they approached Chris Grace, an Animation Director for S4C, the Welsh-language television channel, and he commissioned the Cardiff-based illustrator and writer Rob Lee to develop characters and storylines aimed at children aged from three to five. Lee located Sam and his friends in the Welsh village of Pontypandy, Caerphilly. Originally, Sam starred in a 1986 stop-motion, puppet animation series on S4C. Programmes (each 10 minutes long) about him were also screened on BBC Television for the next 15 years, and five million tie-in books and one million videos were sold in Britain. A team of graphic artists who constitute The County Studio in Leicestershire later provided illustrations for the books.

The number of media and merchandise associated with Sam is extraordinary. Besides the television animation series and videos, they include annuals, a magazine, audio books, CD ROMs, activity packs, playsuits, pyjamas, T-shirts, puzzles, games, drawing books, backpacks, lamps, stickers and soft toys. In 2001, Gullane Entertainment PLC paid £1.6 million to S4C International to acquire a 50 per cent interest in Sam with the intention of making him even more popular. Fireman Sam television series have been sold to 40 countries around the world from Poland to Korea and translated into over 25 languages. Today, there is even

174 Rob Lee (creator), John Walker & Ian Frampton (producers & directors), *Fireman Sam: The Hero Next Door*, (1987). Front cover of 40 mins VHS video recording. Fireman Sam (character) © 1985 Prism Art & Design Ltd. Film © 1987 S4C – Channel 4 Wales. Video © 1987 BBC Enterprises Ltd

175 Virginia Hunter (writer) & Marge Opitz (illustrator), *Fireman Joe*, (Chicago, IL: Rand McNally & Company, 1959). Junior Elf Book series

an interactive website devoted to Sam and his friends.

A profile of Sam reads:

'The friendly fireman from Pontypandy who's always ready and willing to help. He really is the 'hero next door'. Whatever the jam, the word from Sam is 'No Problem'. He's also a great inventor and likes nothing better than to sit in his inventing shed at the end of the garden creating weird and wonderful machines.'

Sam has also been described as a 'perennial bachelor' and he has exhibited human faults such as forgetfulness. In appearance, Sam is short (small children reach up to his shoulders), wears a yellow helmet and trousers, with a navy blue jacket and dark boots. (This uniform was worn by British firefighters during the 1970s and 1980s.) He has blue eyes, a curly forelock of reddish-brown hair, a blob-like nose, a grin, and a large, round, clean-shaven jaw. His colleagues include senior Fire Officer Steele who has a white helmet and grey moustache, and Fireman 'Elvis' Cridlington who is a notoriously poor cook. Naturally, a cast of other characters from the local community are featured, including several small children. An ace female firefighter – Penny Morris from nearby Newtown Fire Station – was introduced later. She often shows more commonsense that her male counterparts and ends up rescuing them. The Fire Station's bright red fire engine Sam drives is called 'Jupiter'.

Sam and his comrades tackle all kinds of routine tasks but sometimes they have to use their hoses against fire-breathing green dragons. As one might expect, given the early age of the audience for Fireman Sam, the storylines are bland and pleasantly amusing. In the books, the simple outlines and strong flat hues of the cartoon images attract the eye but the artistic quality of the drawing is pedestrian compared to the highly imaginative images created by nineteenth-century children's books illustrators such as Arthur Rackham. A number of drawings are outline-only so that children can take an active part by colouring them in. Indeed, some publications are specifically designed as 'colouring books', witness the *Fireman Coloring Book, A Safety Book*, published by the Samuel Lowe Company of Kenosha, Wisconsin in 1950.

Clearly, Sam is a positive role model for small children to admire and emulate. From the earliest age, therefore, children are taught about the work firefighters perform and are encouraged to regard them as friendly, helpful, reliable, everyday heroes. Fire brigades in Britain have naturally taken advantage of Sam to deliver safety messages to children.

## 176 — 1987

### Jeremy Nicholl, Funeral of Colin Townsley

## 177 — 1882

### W.P., The Fireman's Funeral

Unfortunately, this book would not be complete without at least one image of a firefighter's funeral because fatalities recur in all fire services throughout the world. Such funerals – for famous fire chiefs and officers killed on duty – have taken place since the beginning of the firefighting profession and, since the invention of the

176 Jeremy Nicholl (photographer), *Funeral of Colin Townsley*, (1987). Photo courtesy of Jeremy Nicholl and NI Syndication

177 W.P. (engraver), *The Fireman's Funeral*, (1882). Wood engraving, illustration, 23 x 15 cms, published in The *Illustrated London News*, 23 December 1882, p. 652. Author's collection

camera, have been recorded in thousands of photographs. Engravings depicting them also appeared in nineteenth-century illustrated magazines. For example, one showing the funeral cortège of Thomas George Ashford was printed in *The Illustrated London News* on 23 December 1882. Ashford, a 35–year-old Assistant Officer, was killed by falling masonry during a blaze that destroyed the Alhambra Theatre on 8 December. Captain Shaw and about 270 firemen were among the mourners in the procession that ended in Highgate Cemetery.

Most firefighter funerals involve public processions with coffins draped with flowers or national flags mounted on fire vehicles accompanied by officers marching alongside or behind and lining the route. Members of the public witness these processions and are reminded of the sacrifices made on their behalf.

The grief, respect and solemnity associated with these occasions is effectively conveyed by Jeremy Nicholl's news photograph of the funeral cortège of Colin Townsley on its way to the Lewisham Crematorium that took place on 27 November 1987. The closeness and unity of the firefighters in their dress uniforms standing to attention is enhanced by the spatial compression of the depth of field resulting from the long-lens photography.

Townsley (aka 'Tonka' and 'The Guv'nor'), a tall, handsome and fit 45–year-old man, married with children, was Station Officer of the Red Watch crew at the Soho Fire Station in central London. On the evening of 18 November 1987, he travelled on a pump ladder vehicle to a horrendous fire in London's busiest Underground Station at King's Cross. As Martin Lloyd-Elliott has pointed out, fires in underground transport systems are particular dangerous and hard to combat because of the confined spaces, lack of light and ventilation, and the extremely high temperatures that occur. At King's Cross, after wooden escalators caught alight, the temperature reached 1000 degrees Celsius. Inside the ticket hall, Townsley was directing his crew and urging passengers to escape when a flashover occurred. As he fled, he saw a woman struck down by the fierce blast of flames, smoke and heat, and he stopped to help her towards an exit but yards from it he collapsed. His men brought him out but he subsequently died from his injuries. (In total, the inferno caused 31 fatalities.) Townsley was killed because of his unselfish act in trying to save another's life. Posthumously, the Queen awarded him the George Medal for gallantry. (In 2002, to mark the 15th anniversary of the fire, BBC Learning produced a 29–min documentary as one of a series of videos about disasters.)

Townsley's funeral was a 'full-brigade' affair. His coffin, covered by a Union Jack flag and with a white helmet and medals on top, rested on a platform beneath a turntable ladder on a fire vehicle. The latter was dressed with wreaths and six bareheaded pallbearers stood alongside the bier. The affection and esteem in which Red Watch held him was demonstrated by the floral tribute of white chrysanthemums spelling the word 'GUV' mounted above the cab of the red-painted vehicle/hearse. 3000 fire officers representing every brigade in Britain lined the route.

In the aftermath of 9/11, funerals and memorial services for 343 New York firefighters took place week after week. In 300 cases, no remains had been recovered and so there was no body for the families to bury or cremate. The sheer number of services – on one day, 27 occurred! – put a tremendous emotional strain on the 60 bagpipers who had sworn to play at every service and the fire chiefs who wanted to attend as many ceremonies as possible. Dennis Smith, a retired firefighter and acclaimed author of books on firefighting, has provided moving descriptions of the services he attended in his book *Report from Ground Zero* (2002) and, as we shall see shortly, a play and film were produced about the difficulty of writing eulogies to the fallen.

## 178 — 1991

### Tom Corbin, Firefighters' Fountain

## 179

### Detail of the third, standing figure.

179 Detail of the third, standing figure. Photo courtesy of Corbin Bronze Ltd

178 Tom Corbin, *Firefighters' Fountain*, (1991). Memorial/public sculpture, two bronze statues, 2.4 m high, 24 m diameter fountain basin. Kansas City, MO: West 31st Street and Broadway, Penn Valley Park. Photo: Kansas City Public Library

Kansas City, Missouri is known as 'the City of fountains' because it has so many. A large and spectacular circular fountain holding 76,000 gallons of water was designed by Byron Bash and erected in 1991 at a cost of $1.18 million. Funds were raised via private donations and from Kansas City's Department of Parks & Recreation. The basin has two rings of 48 real brass fire nozzles squirting water towards the centre where there are two, over-life-size, realistic, bronze figures of firefighters. One stands and points while the other kneels to control a hosepipe nozzle. Tom Corbin, an artist who was commissioned after winning a competition that attracted 50 applicants, created them. To model the two figures, Corbin posed in front of a mirror dressed in a bunker coat and a helmet. The Fountain was dedicated at a ceremony held on 6 October 1991.

Near the Fountain is a circular memorial terrace with a semi-circle of pillars listing the names of over 80 fallen firefighters. A third, bareheaded, African-American, life-size, firefighter figure stands on the terrace with his head bowed and eyes lowered as in the act of sorrow and mourning. With this figure, Corbin memorialised six local firefighters who had recently perished in a massive explosion.

Corbin (b. 1954) grew up in Centerville, Ohio but now resides in Kansas City. He studied business and art at Miami University, Ohio, and worked in advertising before becoming a sculptor and furniture designer. His interest in sculpture began in 1982 and he received tuition from an established bronze sculptor. His company Corbin Bronze Ltd., which undertakes both private and public commissions, and issues limited edition bronzes of sculptures and furniture, has proved commercially successful.

## 180 1991

### Phil Caruso, Actor Kurt Russell in Backdraft

Universal Studios' fast-paced action, drama and mystery movie is perhaps the most ambitious and elaborate that has been made about firefighting. Gregory Widen, its screenwriter, had personal experience of the subject because before attending film school he had worked as a firefighter in Laguna Beach, California. In fact, he had witnessed a colleague being injured by a 'backdraft'. This dangerous and feared phenomenon occurs when fire consumes all the oxygen in an enclosed space, leaving only unburned natural gases. When firefighters break in, the reintroduction of oxygen triggers a violent explosion.

Ron Howard directed *Backdraft* and its stars were Kurt Russell, William Baldwin, Robert De Niro, Scott Glenn and Donald Sutherland. Russell and Baldwin play two feuding brothers – Lieutenant Stephen 'Bull' McCaffrey (the elder, an experienced but reckless fireman) and Brian McCaffrey (the younger, a rookie or 'probie'). They are second-generation firemen because their father died a hero's death 20 years before in 1971. A prologue shows their father (also played by Russell) rescuing a girl from a house fire but then being killed in a fireball explosion that his son Brian witnesses. Thus, the film opens with the 'saved' motif and it is repeated later. The plot is truthful about firefighting being, in many instances, a family tradition.

*Backdraft* is set and was shot in Chicago, a city whose firefighters have a reputation for being particularly tough and aggressive (in their assaults on fires). In Chicago's Chinatown, the brothers and their friends in Engine Company No. 17 fight blazes some of which are set by an arsonist to generate backdrafts designed to kill certain individuals but then to blow themselves out. De Niro plays Lieutenant Donald 'Shadow' Rimgale a veteran fire investigator who is tracking the arsonist. (Rimgale was the name of a real fire investigator in Chicago.) Bill Cosgrove from the Office of Fire Investigation was hired as De Niro's technical assistant and later wrote a book about their encounter and subsequent friendship entitled *Robert De Niro and the Fireman* (1997). The arsonist turns out to be disaffected firefighter John 'Axe' Adcox played by Glenn.

Several real firefighters from different Chicago engine companies played bit parts; for example, Kevin Casey, Richar Lexsee and Cedric Young. Taro Yamasaki later photographed the trio. Chicago fire chiefs also closed streets and organised a mass funeral parade with a bagpipe band that ended at a replica of Leonard Volk's 1864 *Firemen's Monument* in Rosehill Cemetery.

The plot was implausible and overcomplicated: it included a broken marriage and a love affair between Baldwin and his girlfriend played by the sultry actress Jennifer Jason Leigh. Baldwin and Leigh even perform a passionate sex scene on the top of a fire truck! They make love, not on a bed of roses, but on a bed of hoses! The lovers are forced to call a halt when an alarm sounds and the truck leaves the firehouse at speed with its horn blaring. They are still on board when it arrives at the fire. A firefighter unreeling a hosepipe at the rear of the truck is surprised to find it is draped with Leigh's black lace bra. The plot also includes a corrupt Chicago politician and property developers who are cutting the Fire Department's budget and closing firehouses in order to make profits and this is why Adcox set fires to kill them.

What makes *Backdraft* worth watching are the spectacular and terrifying scenes of flames, smoke and explosions. Indeed, fire is depicted as a near-supernatural

180 Phil Caruso (photographer), Actor Kurt Russell in *Backdraft*, (1991). Publicity still for the film.© 1990, Universal City Studios, Inc

force with a personality of its own. Sutherland, who appears in a cameo role as a sinister pyromaniac serving a prison term, calls fire 'the animal'. When professional firefighters saw the film, they judged its fire scenes authentic even though the climactic one is completely over the top. Allen Hall coordinated the special effects and pyrotechnics, and the Chicago Fire Department gave technical advice, loaned surplus clothes and appliances, and stood by for safety's sake. Principal actors received a month's training and spent a night at a Chicago firehouse; during shooting, they experienced real fires, minor burns and smoke inhalation, while camera crews had to wear air tanks and fireproof suits.

On its release in the United States, *Backdraft* grossed $78 million plus $40 million from theatrical rentals. A tie-book written by Kirk Mitchell was also published. The film was then distributed around the globe, issued on video and screened on television; consequently, many millions have seen it. In a survey of films American firefighters liked to watch in their firehouses, it came number one.

A second film with the title *Backdraft*, written and directed by William G. Wagner, appeared in 1998. Firefighting was also its subject but this time it was a historical drama-documentary set in colonial Virginia about the reconstruction of an eighteenth-century fire engine.

181 John W. Mills, *Blitz*, (1991). National Firefighters' Memorial/public sculpture, bronze, 3.3 m high x 2.34 m wide. London: Sermon Lane. Photo (c) 2002 Michael Lawrie

## 181 1991

### John W. Mills, Blitz

'Better late than never' might describe the appearance in 1991 of a sculptural memorial to Britain's auxiliary firefighters who were killed during World War II. Its location – Sermon Lane, near St Paul's in London – is appropriate because fires raged all around the Cathedral in the 1940s and it was remarkable that Wren's baroque building – which acquired a national symbolic value – managed to survive the bombing. British sculptor John W. Mills (b. 1933) created *Blitz* after being commissioned by the Founder Master of the Guild of Firefighters, who wanted a memorial to act as a focus for an annual service of remembrance. Each September, a religious service held in St Giles church is followed by a march accompanied by a ceremonial Fire Service Band. Once the parade reaches the memorial, a wreath-laying ceremony occurs.

Mills, who trained at Hammersmith School of Art from 1947 to 1954 and at the Royal College of Art from 1956 to 1960, has had a long and successful career as a painter, sculptor, art school tutor and writer of books about the techniques of sculpture. In an essay about viewing sculptures, he cited the following desiderata: they should attract interest from afar and from intermediary distances; possess strong silhouettes; appeal to the sense of touch; and scale should be appropriate to their settings, especially out of doors.

His compact but dynamic group of three bronze firemen in action, who are dressed in wartime uniforms and helmets, meets these criteria. Two men face one another in a semi-crouching position. They brace their legs and arms in order to control a branch hosepipe passing between them (the forward thrust of a water jet creates a powerful backward force) while the third figure, a sub-officer, stands erect and looks back over his shoulder. Both his arms are raised and his hands point towards the imagined fire. However, his left arm is outstretched while his right curves in a gesture asking for support; his mouth is also open as he shouts orders. The figures stand on a pile of bricks and rubble, among which is a small relief image of a fire engine. In style, the sculpture continues the tradition of nineteenth-century academic naturalism and World War I memorials such as those by C.S. Jagger, but it is executed with considerable élan.

The composition of *Blitz*, in fact, was based on an earlier 70–cms-high bronze dating from 1984–85, which in turn was inspired by a photograph taken in April 1941 of two hosemen damping down a fire in rubble-strewn Cannon Street close to St Paul's. Mills' father-in-law Cyril Demarne OBE, ex-Chief Officer of West Ham

Fire Brigade and author of *The London Blitz: A Fireman's Tale* and *Our Girls: A Story of the Nation's Wartime Firewomen*, commissioned the bronze, which is now displayed in the Hall of Remembrance of the London Fire Brigade Headquarters in Lambeth. It was Demarne's idea to erect a memorial to his fallen comrades and since he knew Mills personally, he asked him to sculpt one. Mills in turn used Demarne as the model for the sub-officer figure. (Apparently, when giving orders, he gesticulated with both arms.) A Firefighters' Memorial Charitable Trust was established to raise the £120,000 needed and the Meridian Foundry, Peckham, London cast Mills' full-scale models made from polystyrene and plaster.

During the period 1939 to 1945, 1300 men and women from the fire and police services were killed or died from injuries and most of their names are listed alphabetically on bronze plaques fronting the sculpture's 90–cms-high octagonal base. The style of lettering employed on the base is Gill Sans, a Roman-style typeface dating from 1928–32, designed by Eric Gill, that was used in wartime ration books. The names of 23 firewomen who gave their lives are recorded separately on a bronze panel flanked by relief depictions of a Dispatch Rider and an Incident Recorder.

*Blitz* was unveiled on 4 May 1991 by Her Majesty Queen Elizabeth, The Queen Mother (1900–2002), who had toured the bomb-damaged East End of London during World War II to boost morale. On 6 December 1944, she had also reviewed a parade of 1000 firewomen at the Lambeth Headquarters of the London Fire Brigade.

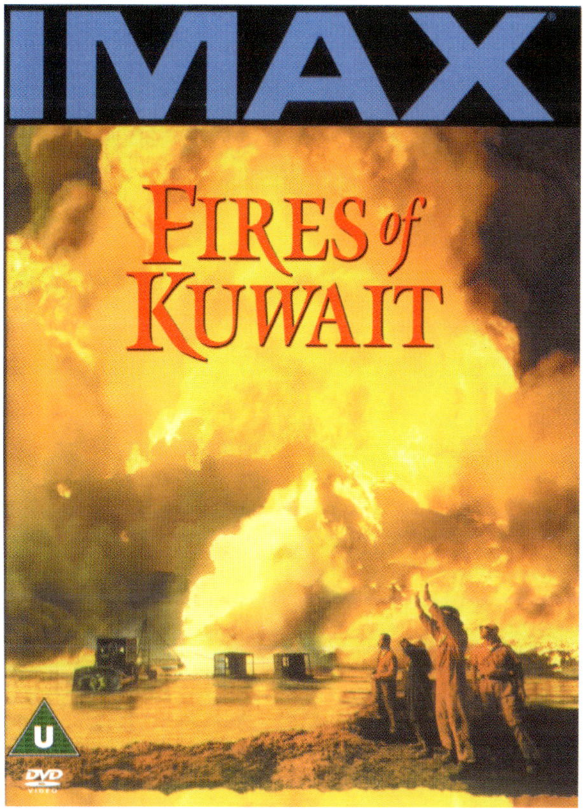


182 David Douglas (director), *Fires of Kuwait*, (1992). Documentary colour film, 36 mins. Toronto, Canada: Black Sun Films Ltd/IMAX Corporation; released on DVD in 2002

## 182 — 1992

## David Douglas, Fires of Kuwait

Arguably, the audio-visual spectacle of fire has never been so vividly portrayed as on the giant screens of IMAX cinemas during shows of the documentary *Fires of Kuwait*. (The term 'IMAX' derives from 'image maximum'. Its film format is ten times larger than 35mm film and the special camera shoots very wide angles. Audiences feel engulfed by the seven-storey-high screen images because they encompass peripheral vision. The sound systems are equally sophisticated.) Black Sun Films Ltd produced this film for the IMAX Corporation of Canada, which was made after Iraqi troops had torched the oil and natural gas fields of Kuwait. David Douglas, a veteran, award winning, natural history cinematographer, filmed and directed it. He and a small crew spent weeks in Kuwait in 1991 while its skies were still black from the smoke from over 600 well fires. Enormous pillars of flame from the burning wellheads provided the only illumination. The desert's surface too was covered with oil and so special roads had to be built to reach the wells. In addition, large lagoons had to be constructed to contain seawater pumped from the Persian Gulf. On-site pumps and hoses sprayed water constantly to cool the ground and prevent machinery from melting. 25 teams of firefighters specialising in oil well control from 14 countries were present and they had imported a huge amount of heavy equipment, some of which had been adapted or invented for the task in hand. A Hungarian team, for example, mounted two fighter aircraft jet engines on a military tank and then used the horizontal thrust of the engines to spray tons of water at fires to drown them. A key lesson of the film was the value of teamwork.

American teams from Texas included the Red Adair Company and Wild Well Control. Joe R. Bowden, who has a university degree in geology, founded the latter in 1975. His son Joe Jr, who has a degree in geography, also works for the company as an engineering technician. Freddy Gebhardt, Vice President of Operations, was team leader in Kuwait and managed to ensure there were no accidents. The company's 18–man team extinguished 137 fires and still found time to assist the IMAX film crew.

The firefighters had to contend not only with high, shifting winds, falling petroleum, blowing dust, sand

and soot, toxic gases, radiant heat from the fires burning at over 2000 degrees Fahrenheit and the extreme noise of the wells but also with landmines and unexploded bombs scattered on the ground. While the film conveys the nightmarish environment in terms of sounds and visuals, it cannot communicate – except through verbal description (the narrator was Hollywood actor Rip Torn) – the horrible smells and the searing temperatures experienced by the fire crews.

Without entering into too much technical detail, Douglas manages to show the various machines and methods used to extinguish the flames, including the use of explosives. Some close-ups were achieved by attaching a camera to the scoop of a mechanical digger while aerial views were shot from a helicopter. There are some hair-raising sequences of firefighters, wearing Nomex suits and heat-resistant long johns, and splattered with oil, working next to wellheads emitting fierce plumes of flames, gas or oil. The dangers and stress they – and the film crew too – faced during their time in Kuwait are palpable. (Some men were killed elsewhere in Kuwait.) They may also experience health problems later from all the polluted air they breathed. Scientists had predicted it would take years to put out the fires, but due to the intensive labour and expertise of the Hellfighters, the task was completed in nine months at a cost of $1.5 billion. Oil well firefighters are highly paid but one of the disadvantages of their profession is that they are away from their homes and families for long periods.

Reviewer Richard H. Curtiss summed up the IMAX film as follows: 'While transporting viewers into the very heart of one of history's greatest man-made disasters, *Fires of Kuwait* also is a riveting depiction of human skill, courage and ingenuity combining to save the planet from the consequences of human folly.'

Of course, Douglas was not the only filmmaker to capture apocalyptic yet sublime scenes, to record the ecological devastation and the determined efforts of the firefighters in Kuwait; much newsreel footage was also shot and noted photographers such as Stéphane Compoint, Peter Menzel and Sebastião Salgado took impressive still pictures. Menzel photographed men from the Houston-based company Boots & Coots International Oil Well Control, founded by Boots Hansen and Coots Matthews in 1978, while Salgado photographed men from the Canadian company SafetyBOSS of Calgary, which makes extensive use of chemical foam to extinguish fires. Jarvis Jackson, SafetyBOSS's team leader, and his crews brought 180 wells under control. Salgado depicted firefighters drenched in oil working close to wellheads, including one man who had been knocked unconscious by a blast of gas. None of the newsreels and still photographs, however, could match the visceral impact of the Douglas documentary viewed in an IMAX cinema.

## 183 — 1993

## Magazine advertisement for Without Warning: Terror in the Towers

At lunchtime on 26 February 1993, a terrorist bomb exploded in the basement car park of New York's World Trade Center. Fortunately, the 107 storeys of the twin towers remained standing but a massive internal crater was created and fires were started, which caused acrid smoke to rise up inside the skyscrapers. Six people were killed, over one thousand injured and 50,000 had to walk down many flights of stairs because power supplies had been disabled. A group of kindergarten children on a field trip was trapped for a time on the observation deck, while another group was trapped in a dark elevator car for five hours. Emergency services were soon on the scene and these included firefighters from Manhattan fire companies such as Rescue 1 and 2.

Soon after the attack, NBC Television decided to com-

183 Magazine advertisement for *Without Warning: Terror in the Towers*, (1993), a telemovie produced by Wilshire Court Productions, Los Angeles, for NBC Television, New York. *TV Guide*, Vol. 41, No. 21, Issue No. 2095, 22 May 1993, published by News America Publications Inc., Radnor, PA

mission Wilshire Court Productions to make a telemovie about it and the rescue efforts. Directed by Alan J. Levi with a script by Stephen Downing and Duane Poole, *Without Warning* was first screened on American television in May 1993 even before the FBI had arrested all the suspects. Writers and researchers interviewed many of those involved and studied documents in order to make the content as accurate as possible. However, one commentator criticised NBC's fact-based emotional dramas because they 'bastardize real events to capitalize on the latest disaster'. The movie was shot on videotape so that it could incorporate documentary footage of exteriors dating from the event itself, while new, fictionalised, interior sequences were filmed, in fact, in Portland, Oregon.

To advertise the telemovie, graphic designers foregrounded the role of firefighters by employing the familiar fireman-with-child rescue motif with the Twin Towers looming in the background. The telemovie recounts several human-interest stories in parallel, which were designed to show that ordinary citizens such as maintenance engineers, office workers and teachers can and do behave like heroes. (In truth, there was also panic and selfishness.) One narrative strand concerns the Rescue 1 firefighters who entered the devastated basement to search for victims. A prominent character is Kevin Shea (the name of real firefighter – born 1959 – who was badly injured) played by George Clooney (b. 1961), a handsome actor who was shortly to star in the television hospital drama *ER*. Poor visibility due to dense smoke causes Shea to fall into the crater and he breaks a leg; his partner Gary Geidel, played by Scott Plank, helps to extricate him. More firefighters arrive with hosepipes to quell the fires caused by burning cars while others help to evacuate the towers and check their 210 elevators.

Given that the telemovie was a rapid response to a recent event, it was reasonably well made and acted but when viewed after 9/11, what was most chilling about it was the *déjà vu* nature of the scenes depicting firefighters and the office workers descending stairs. The real Gary Geidel answered the alarms at the WTC in 1993 and again in 2001. The second time, he did not survive. He died aged 44 leaving a widow and daughter. Lieutenant Shea, who had retired from the NYFD due to his 1993 injuries, immediately went to Ground Zero to help search for Geidel whose body was never found. On 9/11, Rescue 1 lost eleven firefighters. Coincidentally, another Kevin Shea, a New York firefighter aged 34 from Ladder Co. 35, was severely injured during 9/11 and was extricated from the debris by rescuers.

184 Jesse Gardner, *Lieutenant Dennis Mojica with his daughter standing in front of Gardner's portrait of him*, (1994). Photo courtesy of the artist

**184** **1994**

**Jesse Gardner, Lieutenant Dennis Mojica with his daughter standing in front of Gardner's portrait of him**

**185** **1994**

**Jesse Gardner, The artist in New York with a FDNY recruiting poster based on his portrait of African-American firefighter Sheldon Wright**

Jesse J. Gardner (b. 1956) is a skilful professional American artist and designer who trained in New York at the Art Students League (1980–92) and the Parsons School of Design (1986–88). Once a New Yorker, he now lives and works in Philadelphia. Well before 9/11, Gardner portrayed 15 firefighters selected from minority groups in a romantic-realist style. For instance, in 1994 he painted in oils a half-length portrait of Lieutenant Dennis Mojica (of Puerto Rican descent) of Manhattan's Rescue Company No. 1 as one of series entitled 'Unsung Heroes'. The portrait is larger than life and shows Mojica proudly wearing his dress jacket,

185 Jesse Gardner, *The artist in New York with a FDNY recruiting poster based on his portrait of African-American firefighter Sheldon Wright*, (1994). Photo courtesy the artist

white cap and medals. Tragically, Mojica and all ten men of his unit perished on 9/11. In late September 2001, Gardner attended Mojica's funeral service in St Peter's Cathedral and took along the portrait for display inside the Cathedral.

Conscious that firefighting is no longer an all-male profession, Gardner has celebrated female firefighters such as Brenda Berkman who, when she was with Engine Company No. 219, attended the first bomb attack on the World Trade Center on 26 February 1993. Berkman has been an activist for women's rights within the New York fire service since the 1980s. There are 11,000 firefighters in New York but only 27 of them are women. For centuries, firefighting was a strongly masculine culture and so women have struggled to gain acceptance.

Gardner has mounted exhibitions of his portraits, issued prints of them and established a website as a tribute to the spirit of altruism and courage exemplified by his subjects. As the poster depicting African-American firefighter Sheldon Wright in profile holding a hook indicates, the New York Fire Department has also found Gardner's portraits useful for recruitment purposes.

## 186a — 1995

**Joyce K. Killebrew, Storm King 14 Memorial**

## 186b — 2002

**(detail) Bryan Dahlberg, Storm King memorial with a coal seam fire in the background**

## 187 — 1996

**David Nelson, Wildland Firefighters Monument**

## 188 — 2000

**Lawrence Joseph Nowlan, Wildland Firefighters Monument**

On 6 July 1994, a tragedy occurred on Storm King Mountain, in the Rocky Mountains near Glenwood Springs, Colorado. During a forest fire started by lightning, 14 young wildland firefighters, who included four women, perished when a wall of flame caused by severe winds engulfed them. This was in spite of the use of tinfoil protective shelters. One of the victims – Terri Ann Hagen – was a member of the Onondaga tribe of the Iroquois Nation. The fire was later named 'The South Canyon Fire'. Today, a steep mile-long trail up the mountain tracing the route the firefighters took exists as a memorial to the victims. Cement crosses mark the places where their bodies were found.

The deaths on Storm King prompted responses from filmmakers, writers and sculptors. In 1996, a telemovie filmed in Panavision entitled *Smoke Jumpers* (aka *In the Line of Duty: Smoke Jumpers*), directed by Dick Lowry, was made by Patchett Kaufman Entertainment. It focused on the story of Don Mackey, the smokejumper in charge and a father of three, who was one of the victims. Mackey, it seems, could have survived but turned back to help his crew. Unfortunately, the film proved to be a lacklustre account of the romance between Don (played by Adam Baldwin) and his wife René (played by Lindsay Frost) set against a background of forest fires. The plot and the scenes of male rivalry, training, parachuting and firefighting were similar to those of the 1952 movie *Red Skies of Montana*, but unlike the latter, there was no happy ending.

A full and critical account of the Storm King incident, written by investigative journalist John N. Maclean – *Fire on the Mountain* – was published in 1999. Lone Wolf Pictures of South Portland, ME, produced a drama-documentary film based on Maclean's book with narration by

Scott Glenn that was premiered on American television – *The History Channel* – on 28 October 2002. A video was also released.

Joyce K. Killebrew, who has a studio in Sedona, Arizona, never trained as a professional artist but she drew and painted landscapes and portraits all her life until, during her 40s, she began to model and carve sculptures. She herself has been a volunteer forest fire-fighter acting as a lookout and so felt a special empathy with the Storm King victims. She decided to create a memorial and her design was accepted by the Glenwood Springs City Council. The latter wanted a realist sculpture but with figures that were generic rather than portraits of any of the fallen firefighters. Killebrew settled on three figures, two standing and one crouching. Since they have their backs towards one another, they face in different directions and make a compact group, which signifies the solidarity of firefighters. The figures – two men and one woman – are heavily laden with tools, helmets and equipment. They represent three different types of forest firefighter: a smokejumper (seasonal volunteers who parachute into forests to dig fire lines), a Hotshot (elite, professional firefighters) and a Helitack member (helicopter crews who rescue the injured). The Storm King 14 consisted of nine Hotshots, three smoke-jumpers and two Helitack members. The Memorial was dedicated on 6 July 1995 at a ceremony held in Twin Rivers Park. Fourteen boulders with memorial plaques and 14 trees surround the sculpture and its setting is a peaceful one with mountain slopes in the distance.

186a Joyce K. Killebrew, *Storm King 14 Memorial*, (1995). Memorial/ public sculpture, bronze, life-size on a marble base. Glenwood Springs, Colorado: Twin Rivers Park. Photo courtesy of the artist

186b (detail) Bryan Dahlberg (photographer), *Storm King memorial with a coal seam fire in the background*, (2002). Photo courtesy of Bryan Dahlberg/FEMA

To make examples of her firefighter work available to private collectors and to raise funds for a Storm King 14 Memorial Scholarship Fund, Killebrew produced versions of the sculpture in a Maquette edition of 50, available either in Lost Wax Bronze or Bronze Finished Resin. She also issued limited editions of Cold Cast Bronze figures (31 and 36 cms-high) mounted on walnut bases representing wildland firefighters.

Nine of the Storm King 14 had belonged to the Prineville Hotshots, a crew from Prineville, Oregon. Their grieving parents decided to commission a monument in their memory and to honour all those who have lost, and will lose, their lives fighting forest fires. David Nelson, a sculptor from Marble, Colorado, created the

187 David Nelson, *Wildland Firefighters Monument*, (1996). Memorial/public sculpture, bronze, 4 m long, 2.4 m wide, 3.3 high. Prineville, Oregon: Ochoco State Park. Photo courtesy of Charles I. Manetta, Miami

Monument and it was sponsored by STIHL Inc., a company that manufactures chainsaws and other power tools. Again, three bronze figures – one-and-one-quarter life size, two males and one female – laden with tools and equipment are represented but this time they are separate from one another. They cluster around the stump of a blasted tree and are shown working with an axe, digging tool and a chainsaw. All rest on a low stone base. The sculpture is entitled *Wildland Firefighters Monument* and is located in the verdant setting of Ochoco State Park, near Prineville. The dedication ceremony took place on 15 June 1996. A small replica of the Monument mounted on a wooden base was given to STIHL Inc., which uses an image of it on its website. This image also serves as a symbol or icon of the National Forestry Heroism Award, which STIHL makes annually.

Lawrence Joseph Nowlan Jr (b. 1966), a sculptor from Philadelphia but now based in Windsor, Vermont, completed a third sculpture commemorating wildland firefighters in 2000. Nowlan first became fascinated by sculpture while studying at Millersville University in Pennsylvania, which had a bronze foundry. After working in a design company for several years, he resumed learning to sculpt in 1994 at the Pennsylvania Academy of Art and then earned a Masters Degree in Fine Art at the Graduate School of Figurative Art, New York Academy of Art. Since then, he has fulfilled many commissions for figurative sculptures from both corporations and individuals. He and his brother Joseph founded the American Sculpture Association to further their chosen profession.

Nowlan's *Wildland Firefighters Monument* is located at the National Interagency Fire Center (NIFC), America's management and logistical support centre for wildland firefighting, near Boise Airport, Idaho; consequently, it is a national rather than a local memorial. It consists of three, over-life-size, realistic bronze figures placed several yards apart to form a triangle that are establishing a fireline on a ridge with the aid of a chainsaw, a shovel and a Pulaski. (Edward Pulaski, a Forest Ranger, invented the Pulaski, which is a half-axe/half-hoe tool unique to wildland firefighting used for chopping, grubbing and trenching.) The Monument is situated in a landscaped area, which is an acre in size and includes grass, wildflowers, shrubs, birch groves, walkways, a waterfall and plaques listing the names of over 600 wildland firefighters who have died on duty since 1910. Nowlan began work on the Monument in spring 1996 and as part of his research joined firefighters at the Bull Fire in eastern Oregon in the summer of that year.

Vickie Minor of Northwest Contractors, Boise and Executive Director of the World Wildland Firefighter Foundation originated the idea for the Monument. Having visited the Vietnam War Memorial in Washington DC, she thought such monuments had a soothing effect on the psyche. The funds needed – $100,000 – were raised from many businesses and organisations, and by the efforts of volunteer firefighters. The Monument was unveiled at a ceremony held on 25 May 2000, which hundreds attended. Aircraft used for smokejumping flew overhead and dropped purple streamers.

A 76–cms-high model of one of Nowlan's figures is

188 Lawrence Joseph Nowlan, *Wildland Firefighters Monument*, (2000). Memorial/public sculpture, bronze, one of three figures 2.4 m high. Boise, Idaho: grounds of the National Interagency Fire Center. Photo courtesy of the artist

189 Stefan Gec, *Natural History* (1995). Scanachrome photographs enlarged, mounted on roof of Pilgrim Street Fire Station, Newcastle-upon-Tyne. Photo courtesy of the artist

given by the NIFC to winners of the annual Pulaski Award, that is, to the most outstanding Hotshot crew in the country.

All three sculptures discussed above are very similar in conception, style and spirit. They are figurative/realist because this is the idiom most appreciated by the communities for which they were created.

## 189 — 1995

### Stefan Gec, Natural History

Firefighters in different countries face common dangers and so they manifest international solidarity. For instance, firefighters from Newcastle-upon-Tyne visited their counterparts in the Ukraine following the 1986 nuclear power station disaster of Chernobyl. Nine years after the event, Stefan Gec (b. 1958), a British artist specialising in public art projects whose father was Ukrainian, mounted a display of huge portraits of Ukrainian firefighters on the roof of Newcastle's Pilgrim Street Fire Station. The Scanachrome photographs, entitled *Natural History* (1995), were enlargements of passport-like images of the first six Ukrainians who died because of their exposure to radioactivity soon after fighting the blaze at Chernobyl. Gec had in mind the celebratory images of Soviet politicians previously found on political banners but his simplified, black-and-white portraits also resembled some of Andy Warhol's. Later, in 1998, *Natural History* was mounted on the roof of the Ukrainian Education and Cultural Centre in Winnipeg, Canada. Gec's aims were to provide a memorial and to unite the Soviet and Canadian fire brigades 'through a public display of sympathy and reverence'.

## 190 — 1995

### Charles H. Porter IV, Firefighter Chris Fields carrying baby Baylee Almon

## 191 — 4th century BC

### Lysippus (?) Silenus holding infant Bacchus or Dionysus

Just after 9 am on 19 April 1995, a massive terrorist bomb exploded at the Alfred P. Murrah Federal Building in Oklahoma City. Half of the nine-storey building was destroyed and several nearby buildings were damaged. Hundreds were injured and 149 adults and 19 children were killed. (A day-care centre entitled 'America's Kids' had been located on the second floor.) The explosion was heard for miles and so police, fire crews and rescue units rushed to the scene and began a search and rescue effort that was to last 16 days. They worked in dangerous conditions and under constant scrutiny from anxious relatives and the world's media.

Assistant Fire Chief Jon Hansen later described these events in an emotionally charged book entitled *Oklahoma Rescue* (1995). He recalls a poignant moment when a child's red fire truck was recovered from the rubble and

190 Charles H. Porter IV (photographer), *Firefighter Chris Fields carrying baby Baylee Almon*, (1995). Photo (c) Charles H. Porter IV

191 Lysippus? *Silenus holding infant Bacchus or Dionysus*, (4th century BC). Sculpture, marble. Hellenistic or Roman copy. Rome: Pio Clementino Museum, Vatican Museums

reported that two Oklahoma City firefighters lost members of their families in the explosion. What he also records is the tremendous affection, sympathy, support and respect the rescue workers received from the American people.

Chris Fields, a captain from Fire Station No. 5, aged 31 and married with a two-year-old son, was at the scene when police sergeant John Avera emerged from the smoke and ruins and placed 'a critical infant' in his arms. Checks revealed the baby girl – Baylee Almon aged one – was dead. As Fields, wearing a red helmet, tenderly carried Baylee's bloody, dusty and limp body towards an ambulance, he was photographed in colour by two amateur cameramen standing close to one another: Charles H. Porter IV, a 25–year-old bank employee, and 57–year-old Lester 'Bob' LaRue, who worked for a natural gas company. LaRue's photograph earned him $14,000 and made the cover of *Newsweek* magazine (1 May); however, his employer later fired him for refusing to sign over the copyright of the photograph, which had been taken on company time with a company camera. Porter's image was purchased by the Associated Press agency and, in 1996 was awarded a Pulitzer Prize for spot news photography. Both photographs were reproduced numerous times in the world's press because they visually encapsulated the moment and tragedy of the Oklahoma bombing. Aren Almon, the infant's mother, thought her daughter had come to represent everyone who died in the building and to symbolise the innocence of the victims.

While these heart-wrenching icons were contemporary, at the same time they repeated the 'saved' motif that was over 100 years old in terms of firefighting iconography. Furthermore, if one examines the history of representations of a man tenderly holding an infant then one can find examples in Greek sculpture of the Hellenistic age (323–150 BC) such as the carved marble figure of Silenus holding the infant god Bacchus, which has been attributed to the sculptor Lysippus. Thus, the idea that Fields represented the post-feminism 'new man' – one willing to express tenderness and emotion, which tough firemen are in fact prepared to do – is contradicted by the venerable age of this image.

Because of Porter's photograph, Aren Almon and Captain Fields became friends. Furthermore, sympathisers touched by the image sent her $50,000. Soon the firefighter-with-child image was being used on commemorative coins, lapel pins, plaster statuettes, T-shirts and phone cards. However, Almon was upset by the rapid commercial exploitation of her daughter's death. The photographs also affected Fields' life because of the intense media interest that followed. Some of his colleagues were resentful that the media had singled him out when the rescue had been a team effort. Fields himself was suspicious of the publicity because he did not think of himself as a hero or superman and fully expected his celebrity to be short-lived.

Both Almon and Fields subsequently helped a private foundation called 'Protecting People First', whose mission is to raise public awareness of the danger of flying glass that injures so many in explosions and to campaign for the use of glass with protective coatings. The foundation employs a graphic version of the firefighter-with-child image as a logo and supplies a plaque with the image to institutions that install shatter-resistant glass. Almon and Fields hope that one day the image will signify safety rather than tragedy.

## 192 1996

### Joe Quesada & Jimmy Palmiotti, Ash: The Fire Within, Vol. 1

Comic books have been popular with young people in the United States since they emerged during the 1930s because of their cheapness, action-packed stories and colourful dynamic illustrations. They also appeal because they star costumed duel-personality heroes who usually possess superhuman powers, which they need to triumph over evil forces. The most famous of such characters, of course, is Kent Clark/Superman invented by Joseph Shuster and Jerry Siegel in 1938. Superman in turn can be traced back to Friedrich Nietzsche's concept of the *Übermensch*.

Joe Quesada, a prolific award-winning American comic book illustrator and writer, was born in New York City in 1962 and studied illustration at the New York School of Visual Arts. In 2000, he was appointed Editor-in-Chief at Marvel Comics. Earlier, in 1994, he and Jimmy Palmiotti (b. 1961, New York City), another talented illustrator, and Laurie Bradach founded Event Comics of Ottawa, Illinois. Their flagship character was Ashley Quinn, a firefighter who magically transforms into a fiery urban superhero called Ash and who battles against Adam, an alien determined to destroy the Earth. As can be seen from the cover of the first issue of *Ash*, the transformation is implied by superimposing a whirling muscular figure dressed in a strange blue-and-yellow costume over a realistic portrayal of the head and shoulders of a human firefighter wearing a helmet and carrying an axe. This comic book, printed on glossy paper with rich full-colour illustrations, proved an immediate bestseller. Steven Spielberg of DreamWorks optioned it for a possible animated feature film.

*Ash*'s first issue was dedicated to the Fire Departments of Charlotte, North Carolina, Poway and San Diego, California 'for their selfless work on behalf of burn victims and survivors'. Event Comics has been closely involved with several fire departments and has helped to raise funds for charities such as the Firefighters' Burned Children Fund. The firefighters in turn have advised the illustrators to ensure their drawings of firefighting and equipment are accurate.

192 Joe Quesada (pencil and story) & Jimmy Palmiotti (inks and story), *Ash: The Fire Within*, Vol. 1, No. 1, September 1996. Cover of comic Book, 17 x 26 cms. Ottawa, IL: Event Comics, Laurie Bradach Publisher. (c) Invisible College, Inc. Author's collection

## 193 1998

### Timothy A. Maslyn, The Save

In 1993, Timothy A. Maslyn, an American artist born in 1957, established Maslyn Studios in Belle Mead, New Jersey in order to produce realistic figurative sculptures and portraits modelled in clay and then cast in bronze. Maslyn was trained by Joseph Brown (1909–85) in the late 1970s and later became his assistant. Brown was an ex-boxer and Princeton University lecturer in the Creative Arts responsible for many public sculptures of sportsmen and athletes in the Philadelphia area.

Maslyn's sculpture of a fireman carefully cradling a small child in his arms was commissioned by the New Jersey Firemen's Association to honour both paid and volunteer firefighters and is appropriately situated in the grounds of a Home for aged and ill firefighters. (The Home, established in 1898, also contains a museum of fire service memorabilia.) The statue is located just inside the main entrance and is set against a background of lawns and trees.

The grave and dignified bronze figure depicted in the act of stepping forward, rests on a marble base that has a dedication and an incised Maltese Cross. As the sculpture's title indicates, this work is yet another instance of the 'saved' motif that dates back to the mid-nineteenth century and the composition is very similar to the two Oklahoma City photographs of 1995 discussed earlier.

193 Timothy A. Maslyn, *The Save*, (1998). Sculpture, bronze, 2 m high, mounted on a marble base. New Jersey Firemen's Home, Lathrop Ave, Boonton, New Jersey. Photo courtesy of Tim Maslyn

## 194 1998

### Hai Ying Wu, Fallen Firefighters Memorial

This public sculpture, set among a grove of trees at Occidental Avenue South and South Main Street, Seattle, near the Headquarters of the City's Fire Department was commissioned in 1995 following the deaths of four firefighters during the Mary Pang Warehouse blaze in Seattle's Chinatown/International district. (The fire occurred on 5 January 1995 and was caused by arson.) However, the Memorial also honours over 30 firefighters who have died on duty since the Fire Department was established in 1889. The sculpture was unveiled and dedicated at a ceremony held on 6 June 1998 attended by Mayor Paul Schell, Fire Chief James Sewell, firefighters and their families. The event occurred during the annual Fire Festival, which marks the Great Seattle Fire of 1889 that destroyed much of the city.

Don Kennedy, a wealthy local property owner, and Wesley Goss, Battalion Chief, thought a memorial was needed and they asked Norman Taylor, Chair of the University of Washington's School of Art sculpture programme, if his students could help. In fact, over a dozen of the latter executed much of the work on the sculpture under Taylor's direction. One hundred bronze sections were produced, many of which were cast from actual firefighters' uniforms, and then welded together. Funds for the sculpture, which cost $225,000, were raised by the local community and from the sale of artworks

194 Hai Ying Wu, *Fallen Firefighters Memorial*, (1998). Public sculpture, four bronze life-sized figures plus granite slabs. Seattle: Occidental Park, Pioneer Square. Photo courtesy of the artist

donated by members of the Overseas Chinese Artists Association of which Hai Ying Wu, the principal creator of the Memorial, is a member.

Wu (b. 1962) had been trained as a public arts sculptor in China and graduated from the Sichuan Fine Art Academy in 1986. Having survived the 1989 Tiananmen Square massacre, he moved to the United States in the same year. He was studying for his Master in Fine Art qualification at the School of Art, Washington when his design won the competition for the commission. The fact that the fire happened in Chinatown and a Chinese artist was selected to sculpt it was coincidental.

The Memorial consists of four statues of anonymous firefighters standing, crouching and kneeling next to a ruined wall and broken beams. (Viewers are left to imagine the inferno that confronts them.) Wu has explained that he wanted the granite beams – on which the names of the fallen are inscribed – and the overall pyramidal composition to signify the strength and toughness he associates with firefighters. The figures wear breathing masks to hide their individual identity, gender and race, and carry such equipment as axes, chainsaws, compressed air bottles, fireflies (that is, electronic signal devices), hose nozzles and shingle rakes. In Communist China, there is a tradition of dramatic realism in sculpture – witness the *Rent Collection Courtyard* ensemble of 114 life-sized clay figures created by a collective of amateur and professional artists from Szechuan province in 1965 – that Wu was surely familiar with. The Seattle Memorial is different from earlier ones in depicting a scene of firefighters in action and having no plinth. The absence of a plinth means that the statues and viewers are on the same level. Furthermore, the separation of the figures enables spectators to move among them; this endows the sculpture with a strong sense of immediacy.

During 1996 and 1997, Wu created firefighter sculptures for the Washington cities of Renton and Redmond, and in 1999, following the success of his Seattle Memorial, he was commissioned to make a further firefighter relief sculpture entitled *Rescue* for a fire station in Mercer Island, Washington.

## 195 — 1998

### Robert J. Eccleston, New York State Fallen Firefighters' Memorial

Robert J. Eccleston has an unusual background for a visual artist. He was born in New York City in 1964 and, during the 1980s, he studied industrial design at the School of Visual & Performing Arts, Syracuse University. After graduating in 1987, he joined the United States' Infantry, became an expert in mountain warfare, and served in several countries around the world. In 1993, Captain Eccleston resigned his commission to devote himself to the art of sculpture. From 1993 to 1998, he undertook an Independent Study in Sculpture programme at Plattsburg State University of New York and gained further experience with such established sculptors as Fritz White and Frank C. Gaylord. His artistic abilities soon gained him awards and important commissions from both private and public patrons. Currently, he maintains a studio and foundry called the 'Cloudsplitter Studio' in the Adirondack Mountains where he generates sculptures of historical figures, wildlife and war memorials.

In October 1997, the Office of General Services of the State of New York commissioned Eccleston to design and construct a memorial to the fallen firefighters, both volunteer and paid, of the State. The proposed site was the Empire State Plaza, in the County seat of Albany, next to the Justice Building and near to a fire engine museum and the headquarters of a Firefighters' Association. Eccleston decided on a figurative sculpture depicting three male firefighters emerging exhausted from a blaze. (The Tallix Foundry in Beacon cast his models in bronze.) The central, bareheaded figure is severely injured and is being dragged to safety by companions on either side. One colleague strives to keep the

195 Robert J. Eccleston, *New York State Fallen Firefighters' Memorial*, (1998). Memorial/public sculpture, bronze, 3 m-high. Empire State Plaza, Albany, New York. Photo: Tom Semeraro, courtesy of Robert J. Eccleston

injured man conscious, while the other takes the lead in seeking a way out. Eccleston's aim was to convey 'the firefighter's indomitable will to survive and the camaraderie needed to overcome adversity'. There is a strong sense of drama and movement as the figures lean forward and the group sculpture, set on a low circular podium, is tightly focused because the men are close together and are supporting one another. Thus, the unity and compassion of firefighters is eloquently and succinctly portrayed. The men's heavy, 1960s' style turnout coats, boots and helmets are rendered in a way that is both bold and detailed, and their faces are modelled with a high degree of expression and realism. The bronze was painted black and then details were burnished. Eccleston describes his style as 'rough texture with highlights. The rough texture represents the seasoned firefighter covered with soot. The smooth highlights depict sweat and water running down the firefighter's face and clothing.' Although this sculpture is conventional and traditional in conception and style, it has been executed with considerable verve. It also has a strong emotional charge.

Architect Jim Jamieson designed the extensive memorial site, which includes granite benches and a grassy area with box-trimmed, Norway, maple trees. Immediately surrounding the sculpture is a circular area made from charcoal and red pavers, which features the image of a Maltese Cross.

At a public ceremony held on 5 October 1998, Governor George E. Pataki – himself the son of a volunteer fireman – unveiled the Memorial, which is dedicated to the firefighters (2241 at the time of writing) who lost their lives while on duty in New York State since 1806. Their names have been sandblasted into the surface of a long grey-granite wall that stands behind the bronze figures. When viewed from the front, this wall forms a light-coloured backdrop against which the darker sculpture is contrasted.

## 196a/b 1998

### Bullard, Two Thermal Images

These two blurred representations of firefighters and a casualty resemble, in their abstracted form, examples of avant-garde art; two stills from a work by a video artist perhaps. However, they are not examples of recent art but pictures taken by a Thermal Imaging Camera (TIC), that is, one that registers infrared radiation rather than visible light and is therefore capable of detecting heat-emitting objects and people hidden from human eyes either by darkness or thick smoke. One of half-a-dozen companies designing and manufacturing these cameras is the American firm Bullard, which was founded by Edward D. Bullard in 1898 in San Francisco. It is now

196a 196b Bullard, *Two Thermal Images*, (1998?). Photos reproduced courtesy of Bullard, Cynthiana, Kentucky

based in Cynthiana, Kentucky and specialises in safety equipment.

Bullard's first TIC was launched in 1998. (Early uses of TICs by fire services date back to the mid 1980s.) It is a battery powered, hand-held (pistol-grip), waterproof device with an inbuilt screen at the back so that firefighters can instantly view, in black-and-white, what it 'sees'. On screen, warm objects appear light and cool objects dark but some units can reverse the white/black feature. (More recent cameras have been mounted on helmets and are capable of colour images.) TICs have increased the speed of searches and facilitated many rescues of unconscious victims and downed firefighters. In addition, they help firefighters to pinpoint the sources of fires and to identify structural problems. A TIC can also be linked, via a wireless video transmitter, to a monitor in a command truck so that fire chiefs can follow what is happening inside a building and direct their crews accordingly. Furthermore, incident commanders can videotape the images for review and training purposes.

Although the TIC is expensive and has some dangers and drawbacks, it has proved an important addition to the firefighter's tool-kit. The daily use of such a sophisticated high-tech device demonstrates that firefighting is not just a manual job dependent on brute strength but a profession requiring technical skills and training. The latter includes interpreting the images the camera generates.

*Thermal Image* is also the title of a first novel by Pat O'Keeffe, a firefighter in the London Fire Brigade, published by Hodder & Stoughton in 2002. It is a thriller about a London firefighter tracking an arsonist.

## 197 1999

### Steven Whyte & Malcolm Talbot, The Spirit of 49

This sculpture, which was unveiled on 26 March 1999, honours those firefighters who gave their lives while serving the County of Staffordshire, in the Midlands of England, and commemorates the 50th anniversary of the founding, in 1949, of the County's Fire & Rescue Service. Following several months of consultation with

197 Steven Whyte & Malcolm Talbot, *The Spirit of 49*, (1999). Memorial/public sculpture, painted Fibreglass, with a brick base, 1.37 m long x 80 cms wide x 1.55 m high. Located outside the Staffordshire Fire & Rescue Service Headquarters, Pirehill House, Pirehill, Stone, near Stoke-on-Trent. Photo by Snaptastic, Stone, Staffordshire

fire officers and officials from the Staffordshire County Council (who commissioned the sculpture), an agreement was reached regarding the final design, which was then rapidly executed in Fibreglass painted to resemble stone by two commercially successful artists – Steven Whyte and Michael Talbot. Both are skilled at portraiture and public monuments, committed to teaching art college courses in figurative sculpture and have been members of the Royal Society of Portrait Sculptors.

Whyte is English but spent much of his youth in Europe. He studied at the Sir Henry Doulton School of Sculpture, Stoke-on-Trent (now the Frink School of Figurative Sculpture) where one of his mentors was the noted sculptress and printmaker Dame Elizabeth Frink RA (1930–93). He is director of Sculptural Solutions Ltd., and maintains studios in East Yorkshire and in Monterey, California.

Talbot ARA (b. 1959), Whyte's collaborator, is a busy figurative sculptor and art college tutor, who trained at the Newcastle School of Art, North Staffordshire University, the Royal Academy Schools and the Sir Henry Doulton School of Sculpture between 1975 and 1988. Frink taught him as well.

This realist, carved-stone style statue is yet another contribution to the 'saved' motif tradition: a crouching firefighter wearing a modern helmet, visor and breathing apparatus cradles an unconscious girl in his arms. Standing behind him is another fireman who, like a guardian angel, provides support. Since he is dressed in the uniform and helmet of the original brigade, it is clear he is a historical figure embodying 'the spirit of 49' that continues to inspire contemporary Staffordshire firefighters. Paul Longmore and Alan Paulinski, two serving fire officers, acted as models for the firemen and Hayley Wilshire, an eight-year-old local girl, posed for the rescued victim.

## 198 — 2000

**Robert Workman, James Crossley, star of Naked Flame**

## 199 — 1999

**Grey Advertising Inc., Five Alarm Jockey ... Let 'em know you're Jockey Pouch**

As we have seen, in many antique portraits of firefighters helmets have figured prominently. More recently, male strippers have used the modern British yellow helmet as a prop in their stage acts because all-female audiences find them sexy. Since the early 1990s, good-looking firemen and firewomen have been willing to pose semi-naked for calendars, usually described as 'saucy', sold to benefit fire service and other charities. In 2000, Peter Benedict was inspired by such a calendar to write and direct a farce entitled *Naked Flame*, which has since proved popular with theatre audiences throughout Britain.

Benedict researched his comedy in several London fire stations and was advised by the serving firefighters responsible for the *Firefighters Calendar*. (Men at Colchester Fire Station conceived the latter in 1993. It was such a success, they established the company UK Firefighter Publications Ltd.) Audiences mostly consist of middle-aged women enjoying hen nights but a three-quarter-length publicity portrait of actor James Crossley tightly gripping a hose nozzle set against a blood-red background, taken by *Gay News* photographer Robert Workman for the farce, must surely have appealed to homosexuals too.

Glamour calendars similar to the British one are produced in countries such as Australia, Canada, New Zealand and the United States. Some focus on men, some on women and others feature both genders. (Calendars from Worcester, Massachusetts and New

198 Robert Workman (photographer), *James Crossley, star of Naked Flame*, (2000). Photo courtesy of Guy Chapman Associates, London

York City employ more realistic, documentary-type photographs.) Sensual images of firefighters have also appeared in advertisements, diaries, posters and postcards. In 1998, for example, Jarrold Publishing of Norwich issued a set of 30 postcards with all-male images, which had previously appeared in editions of the *Firefighters Calendar*. In these photographs, the emphasis is on muscular arms and torsos, men taking showers and playing with jets of water, grasping axes, helmets, nozzles and sliding poles. In short, the iconography is borrowed unashamedly from gay erotic magazines and still photographs. Similarly, the imagery of female firefighters' calendars and posters, such as those issued by Firechix.com of Canada, reprises the pictorial clichés of pin-ups and soft-porn magazines.

A notoriously eye-catching and 'steamy' 1999 advertisement for 'Jockey' briefs devised by the art directors Ron Castillo and Kira Shalom, and copywriter Jan Egan of the Grey Advertising Inc., agency of New York depicted five, real, male firefighters from Dallas, Texas with their pants around their ankles. (In return for the firefighters posing on 20 November 1998, Jockey International donated $5000 to the Parkland Burns Center in Dallas.) Actually, when firefighters sleep in firehouses they often leave their pants and boots in this position so that they can step into them quickly. One

199 Grey Advertising Inc., *Five Alarm Jockey ... Let 'em know you're Jockey Pouch*, (1999). Magazine advertisement and poster. © 1999 Jockey International Inc., Kenosha, Wisconsin

man holds an axe, another a hook, while a third cradles a Dalmatian. This breed of dog – white with black spots – has been closely associated with firefighting because Dalmatians used to escort horse-drawn fire engines to prevent other dogs interfering. They also used to guard horses and firehouses.

This was not the first time Jockey had employed images of firemen in its advertisements. One published in a 1947 American magazine had an illustration of two nozzle men climbing a ladder plus a photograph of a man in his underwear. The copywriter's pitch was that a fireman's uniform was functional just like Jockey underwear.

## 200 — 2001

### George Danhires, Firefighters' Memorial

Hamilton, Ohio was once known as a significant manufacturing centre. Since 2000, however, it has been called 'The City of Sculpture' because it has so many public sculptures especially in the 40–acre Pyramid Hill Sculpture Park. In 2001, Hamilton added a firefighters' monument, which is located in Memorial Park near the Great Miami River. Two years earlier, the Memorial Committee of the Butler County Fire Chiefs Association began to raise the $35,000 needed to fund a bronze sculpture to commemorate Butler County firefighters and emergency medical service personnel who had died on duty since 1912. The fire chiefs were much impressed by the fine detail they observed in a wax model shown to them by the figurative artist George Danhires of Kent, Ohio and so he was commissioned to create the statue.

Danhires was born in Pittsburgh in 1942 and received a Master in Fine Art award from Ohio University, Athens, in 1974. Since then, he has exhibited both oil portraits and sculptures, and produced many public works of art. In his art, Danhires is keen to pay tribute to ordinary people and community values.

A bronze firefighter dressed in full gear kneels on one knee and cradles an infant tenderly in his arms; hence, the sculpture repeats the by now familiar 'saved' motif and many viewers will be reminded of the photographs taken of Chris Fields and Baylee Almon in Oklahoma City in 1995. Lieutenant Kelly Mayer, a firefighter from the Fire Department of West Chester Township, served as the model for Danhires' main figure.

Like so many other American sculptors creating firefighter memorials, Danhires favours a detailed, naturalistic style of representation, which nevertheless is capable of conveying strong feelings. The statue was erected on a low stone pedestal with a plaque listing the names of the fallen. Schumacher-Dugan Construction of West Chester Township built the base and organised the installation. The Memorial was dedicated at a ceremony held on 25 March 2001.

200 George Danhires, *Firefighters' Memorial*, (2001). Memorial/public sculpture, bronze, life-size. Hamilton, Ohio: Monument Ave, Memorial Park. Photo: courtesy of the City of Hamilton

## 201 — 2001

### Kenny Hunter, Citizen Firefighter

At the junction of Gordon Street and Hope Street in Glasgow, near the entrance to a railway station, stands an over life-size bronze statue entitled *Citizen Firefighter*. It was commissioned by the Strathclyde Fire Brigade, sculpted by the Scottish artist Kenny Hunter (b. 1962), and unveiled on 17 June 2001. The idea for a statue was first mooted in 1998 and two years later, a group of firefighters began raising the £45,000 needed and started to search for a suitable artist. The statue commemorates Scottish firefighters who have died on duty. Hunter researched his subject by attending incidents with a crew from Glasgow's Maryhill Fire Station.

The commissioning firefighters decided against a sculpture embodying 'pseudo-heroic postures' and Hunter himself was sceptical of the grandiose rhetoric of historic urban monuments. His solution involved a combination of some traditional and some modern characteristics. For instance, the figure is still and static like Victorian ones perched on plinths but his stands on a modest podium so that it is approachable and can be touched by the public. As Robert D. Coke, an Assistant Firemaster, put it: 'It is not on a high pedestal and neither should it be; like citizen firefighter's real life counterparts it is an integral part of everyday life.' Ray MacKenzie has also commented: 'It does not perpetuate the Victorian cult of personal greatness by casting the subject in the role of a demi-god ... neither a hero nor an anti-hero, Hunter's firefighter is presented in a posture of disciplined readiness, an ordinary mortal who just happens to have a particularly important job to do.'

Hunter's figure is recognisable because it represents a contemporary firefighter dressed head-to-toe in black protective clothing with helmet, gloves, boots, and heavy oxygen tanks. However, the artist adopted a reductive approach to form to avoid detailed naturalism and for aesthetic, tactile reasons. 'Subtraction,' Hunter maintains, 'peels away pathos.'

The mask of a breathing apparatus conceals the figure's face because Hunter wanted to stress 'the generic or quotidian aspect of the figure' and to 'rekindle the onlooker's reflexes', to encourage spectator participation when they wondered about the person inside. In Coke's view, the statue is 'a lasting and important commemorative figure, which epitomises the personality, integrity and demeanour of the modern "firefighter", whether that individual be female, male, Caucasian or of ethnic origin'.

This public sculpture is undoubtedly arresting but some viewers may find it rather sinister because the fire officer seems imprisoned by all the heavy clothing. (MacKenzie was even reminded of the 'carapace of protective technology' associated with astronauts and deep-sea divers, while Francis McKee discerned 'epic references to a Star Wars storm trooper'.) One American artist who has painted humanistic portraits of firefighters finds the statue 'morbid and alienating'. Nonetheless, those who commissioned it are pleased with it and employ it as the focus for memorial ceremonies. It serves to remind the public of the vital service firefighters at home and abroad provide. Many citizens of Glasgow and visitors to the city also seem to like it because it has become a meeting point and a tourist attraction. Hunter's statue is one of the few sculptures of a firefighter that has a 48–page booklet of essays devoted to it (see Bibliography).

201 Kenny Hunter, *Citizen Firefighter*, (2001). Memorial/public sculpture, bronze with granite base, 2 m high. Glasgow: Gordon Street. Photo courtesy of glasgowsculpture.com

## 202 — 2002

**Butterton, New York Firefighter Louie Cacchioli stands in front of a Polaroid photo of himself by Joe McNally & Mark Sobczak in the Faces of Ground Zero exhibition, Royal Exchange, London**

## 203 — 2002

**Richard Kron, Ground Zero**

## 204 — 2002

**Linda Koast, The Phoenix: Three New York Firemen raising the American Flag at Ground Zero**

New York is probably the media capital of the world and at any one time, there are hundreds of photographers working there. Naturally, they responded to the disaster of 9/11 with alacrity. News photographers such as Christophe Agou, David Gamble, Shannon Stapleton, David. C. Turnley and Dennis van Tine, and even fleeing office workers such as John Labriola (who had a digital camera with him) documented the desperate efforts the NYFD made to rescue people from the Twin Towers;

202 Butterton, Andy (photographer), *New York Firefighter Louie Cacchioli stands in front of a Polaroid photo of himself by Joe McNally & Mark Sobczak in the Faces of Ground Zero exhibition, Royal Exchange, London*, (2002). © PA Photos Ltd 2004

consequently, there are dramatic shots of firefighters in action both inside the buildings and outside.

Furthermore, the NYFD has its own Photo Unit headed by Lieutenant Richard Smiouskas, which includes three firefighters. The Unit was established in the 1940s in order to document the daily work of the Department, fire scenes, dedication and funeral ceremonies. In the immediate aftermath of 9/11, the Photo Unit took aerial shots from helicopters to map out the rescue effort.

Naturally, the Unit and other news photographers recorded firefighters searching the mound of debris known as Ground Zero for survivors and bodies. Exhibitions of their photographs soon followed. Joel Meyerowitz's images, for instance, were shown at the Museum of London during March and April 2002. Earlier, in January 2002, an exhibition entitled *Faces of Ground Zero* was held in Vanderbilt Hall, Grand Central Terminal, New York and in March a version of it was shown at the Royal Exchange in London, (it then travelled across America). The show consisted of full-colour, life-size portraits of rescue workers, including firefighters such as Louie Cacchioli of Engine No 47 who narrowly missed being killed.

*Life* photographer Joe McNally, assisted by Mark Sobczak, took the photographs using the world's largest Polaroid camera. McNally has recalled that the mood in the studio was one 'of acceptance and healing' not anger. The portraits were serious, dignified and straightforward: frontal shots of standing figures dressed in their protective clothing set against plain white backdrops. (The composition recalled Edouard Manet's 1866 painting *Le Fifre*.) The photographs, which were intended to capture the pain, sense of loss and the bravery of the survivors, have also been widely reproduced in books and articles about 9/11 (see Bibliography).

At about 4 pm on 11 September, Thomas E. Franklin, a staff photographer for *The Bergen Record*, New Jersey,

203 Richard Kron, *Ground Zero*, (2002). Memorial/public sculpture, painted white oak, figures: life-size. Brockport, New York: Capen Hose Company Fire Station, 237 South Main Street. Photo courtesy Westside News Inc., Spencerport, New York, U.S.A.

204 Linda Koast, *The Phoenix: Three New York Firemen raising the American Flag at Ground Zero*, (2002). Painting, acrylic on canvas, 61 x 122 cms. Photo courtesy of the artist

took a colour, digital, long-lens photograph that was to become famous. It depicted a trio of dusty firemen – William Eisengrein, George Johnson and Daniel McWilliams – raising an American flag, which they had found on the back of a nearby yacht. (Ricky Flores, staff photographer for *The Journal News*, took a similar photograph of the same subject from a higher viewpoint.) The photograph's composition was already very familiar because it echoed that of Joe Rosenthal's famous shot of six American marines raising the Stars and Stripes on Iwo Jima during World War II.

After Franklin's striking photograph appeared in the press, it was reproduced on posters and prints marketed by Artseek.com, on T-shirts and postage stamps. It also inspired plans for a tall bronze monument entitled *Flag Raising at Ground Zero* honouring New York's firefighters. A full-scale, clay prototype designed by StudioEIS – a commercial 3D design and fabrication enterprise founded by Elliott and Ivan Schwartz in 1976 – was unveiled on 21 December 2001. Its style was academic-realist. The completed bronze was due to be erected in Brooklyn in the spring of 2002 at a cost of $180,000. However, the planned sculpture aroused considerable controversy in America because political correctness had prompted a departure from fact: the three firemen who had raised the flag were all Caucasian but the sculpture itself represented one white, one African-American and one Hispanic in order to reflect the racial diversity of NYFD. In January 2002, the sculpture was cancelled because the New Jersey Media Company that owns the rights to the Franklin photograph objected to the proposed alterations in the ethnic identity of the three firefighters. Nevertheless, Avgallery.com is marketing a small, ornamental, hand-painted bonded marble statue of the flag-raising scene executed in a kitsch or folk art style by the Valentino © Collection (an Italian company that produces religious artefacts).

In addition, in the village of Brockport, New York that has a volunteer fire department, a public sculpture based on Franklin's photograph was erected in 2002 outside the Capen Hose Company Fire Station and Firefighting Museum. (In this instance, there does not seem to have been a copyright problem.) It was the idea of D. Scott Warthman, a retired fire chief, who raised funds from the community and built the low, circular stonc basc on which it stands. A local portrait painter and sculptor called Richard Kron, who was born in Germany in 1944 and came to the United States in 1956, created the sculpture, entitled *Ground Zero*. As a sculptor, Kron carves wood with the aid of chainsaws and sanders. His three, life-size figures are made from oak and have been painted to create a realist effect; however, his figures are somewhat wooden in a metaphorical as well as a literal sense. The memorial was dedicated in the presence of a crowd of 3000 on 18 August 2002.

Franklin's image also prompted an acrylic rendition by the American portrait painter Linda Koast after she had obtained a license from the copyright holders. Deeply affected by 9/11, she wanted to contribute a more permanent record of the scene Franklin had captured and to help the bereaved. Limited edition prints of the canvas have been produced for sale to benefit the fallen firefighter's charity The Bravest Fund and Koast wants to donate the painting to the White House or the Smithsonian Museum. Koast is a self-taught artist working very much in the photo-realist, populist manner of Norman Rockwell, an artist she much admires.

There is no doubt that some kind of architectural or sculptural monument to 9/11 will be erected in New York and many proposals and plans have already been put forward. Memorialising the event poses a tremendous challenge to the imagination of architects and fine artists.

According to Beth Nissen of CNN: '... since September 11, firefighters – especially New York City firefighters – have become the subject of a newly pronounced public devotion, reverence, near worship.' Later she described them as 'selfless. Firefighters seem all the more noble because they are motivated by service, not reward. Full-time firefighters are only moderately paid – and 74 per cent of the nation's firefighters are volunteers.'

## 205 2001–02

### Naudet Brothers, 9/11

Live television transmissions, video recordings and films are generally more riveting media than painting, sculpture and still photography because of their high illusionism and 'you are watching it while it happens' quality. On 9/11, Gédéon (b. 1970) and Jules (b. 1973) Naudet, two French brothers and filmmakers, were making a documentary in New York about Tony Benetatos, a 'probie' (probationer) firefighter assigned to Engine Company No. 7 and Ladder Company No. 1, Duane Street Firehouse, lower Manhattan. He had joined the Fire Department because he 'wanted to be a hero' and he tempts fate by hoping for a fire. The Naudet brothers had earlier made friends with James Hanlon, a Bronx-Irish fireman, who became the co-producer of their film along with David Friend and *Vanity Fair*'s editor, Graydon Carter.

Jules was practising his camerawork at a minor incident in a street near the World Trade Center when the terrorist attack began. He recorded the first aircraft crashing into the North Tower (this unique footage was later sold to CNN for $10,000) and followed Battalion Chief Joseph Pfeifer into its lobby to film the chaotic scenes as the command post sent units upstairs to assist and evacuate the office workers. Hence, Jules risked his life along with the firefighters. He experienced panic too when he noticed the worried expressions on the faces of the firefighters themselves.

On 10 March 2002, CBS screened *9/11*, a two-hour television programme compiled from Naudet footage with narration spoken by Hanlon. Film star Robert De Niro – who had played an arson investigator in the movie *Backdraft* (1991) – hosted it even though his presence was superfluous, and around 40 million Americans watched it. In Britain, the programme was shown on BBC 1 television on 11 September 2002. DVD and video editions have since been issued. The first quarter of the programme chronicled the daily chores, jokes and camaraderie among 'the guys' inside the firehouse. This was followed by footage of the attack, the rescue efforts, and of Father Mychal Judge, a fire chaplain at prayer, minutes before he was killed. Self-censorship was imposed because no victims on fire or the dead bodies of those who had fallen or jumped from the top of the Towers were shown although crashes were heard on the soundtrack. This section culminated in Jules' camera recording dust and debris as he ran for his life when the North Tower collapsed. Chief Pfeifer, whose own brother Kevin – another fireman – died, protected Jules by lying on top of him. The shaky, hand-held camera shots

205 Naudet Brothers, 9/11, (2001–02). Television documentary, 120 mins, transmitted on CBS in 2002. Stills from programme courtesy of the Naudet Brothers

taken in conditions of extreme danger were the most compelling. This was participant observation with a vengeance.

Meanwhile, Gédéon was recording the bewildered and horrified reactions of people in the streets around the Twin Towers to an event they could not yet comprehend. He also captured the second aircraft crashing into the South Tower. Once the Naudet brothers were reunited, they continued to film firefighters arriving back at the firehouse (all its members survived) and then their return to Ground Zero to search for anyone trapped. For three weeks, they recorded the rescue efforts and the memorial service for Kevin Pfeifer.

One viewer criticised the programme for being 'sanitized and manipulative' but another thought the catastrophe, horror and courage it documented revealed 'humanity at its most vulnerable and heroic'.

Another television documentary about New York's firefighters was shown in Britain on Channel 5 on 29 August 2002. Entitled *9/11: The Firefighters' Story*, it was made by Anglia Television's 55–year-old producer Paul Berriff, an experienced documentary filmmaker who has been dubbed 'the Indiana Jones of television' because of the dangers he has courted. Like the Naudet brothers, he too had been making a film about New York's firefighters and had been shadowing Bronx Battalion Chief Mike Puzziferro. Berriff and his crew were at the firefighter command centre filming the damaged World Trade Center from below with a small digital camera when the South Tower collapsed. They ran away but Berriff was knocked unconscious for a time by falling debris and his sound recordist was buried under rubble. They were fortunate to survive.

Berriff's hour-long programme, which had a commentary spoken by Puzziferro, focused on the firefighters' experience of the disaster and its aftermath: the rescue efforts and the months of searching for body parts at Ground Zero and attending services for those who had perished. A search for missing fireman Michael Lynch took six months and when it succeeded revealed a macabre and poignant variation on the 'saved' motif: Lynch's remains were intermingled with those a woman, who in the opinion of the medical examiner, he had been carrying or protecting. In another harrowing sequence, Puzziferro inspected rows of fire trucks and other vehicles that had been damaged during 9/11 and in which some of his friends had been crushed to death.

Interviews with survivors and their wives revealed the traumatic and long-term emotional effect on them and their children. In this instance, therefore, the victims were not those heroic firefighters had rescued, but the firefighters themselves. Nevertheless, what the film also showed was a spirit of fortitude and the willingness of the surviving fire companies to continue their daily work of saving lives and protecting property.

206 Matthews International Corporation, *Praying Firefighter*, (2001). Memorial/public sculpture, bronze, 1.8 x 1.8 m. Photo courtesy of www.newyorkled.com

## 206 — 2001

### Matthews International Corporation, Praying Firefighter

In 1850, an English immigrant to the United States named John Dixon Matthews founded the Matthews International Corporation of Pittsburgh, a manufacturing company now noted for its range of 'identification and memorial products and services'. In 2001, its Bronze Division and its sculpture facility in Parma, Italy designed and sculpted a realistic statue of a bareheaded, grieving or praying firefighter in a kneeling position, which had been commissioned by the Fire Fighters' Association of Missouri in 2000. However, since the statue arrived at Kennedy Airport just before 9/11, Matthews and the Missouri firefighters decided to donate it to the New York City Fire Department as a symbol of their loss. Coincidentally, the day the statue was released from Kennedy Airport, the *New York Post* published a full-page photograph of a firefighter at Ground Zero in the same pose as the bronze figure.

For several months, the statue was displayed on a flatbed truck at 44th Street and Eighth Avenue outside the Milford Plaza Hotel close to a firehouse, where it attracted many spectators who contributed American flags, bouquets of flowers, candles, cards and drawings. Also displayed on the truck were two bronze 'feature' plaques (each 122 x 244 cms) one of which had a bas-relief representation of firefighters carrying and accompanying children. Jimmy Donatelli, former owner of the Quintet Monument Company, drove the truck to the Matthews' plant and transported the plaques to New York.

The statue now has a permanent home on a granite slab at the corner of Eighth Avenue and 44th Street. Its installation was paid for by the Milstein family, which owns the Hotel. Bronze plaques and the names of the fallen are to be placed on the Hotel's exterior wall opposite the statue.

The Missouri firefighters were not deprived of a statue because they ordered a duplicate that was erected on a purpose-built memorial site in Kingdom City, Missouri. It forms the centrepiece of a black marble wall recording the names of the fallen. A dedication ceremony was held on 18 May 2002.

Another plaque devised by Matthews in 2002 featured a firefighters' prayer, a Maltese Cross and a representation of the kneeling figure. In October of that year, it was added to a Firefighters' Memorial site near the entrance to Ivy Hill Cemetery, Alexandria, Virginia. This site, created in 1856, was prompted by the deaths of seven firemen killed by collapsing walls during a fire a year earlier. It consists of a tall stone obelisk with a relief figure of a fireman standing in a niche half way up and a fountain with a hose nozzle. The site has since been developed and re-landscaped several times. A 'Circle of Honor', a new 'Friendship Fountain', a bronze relief and five flagpoles have been added. Every October, the Alexandria Fire Department holds solemn memorial and wreath-laying services there.

## 207 — 2001

### Gary Suson, Portrait of Lee Ielpi

Gary Suson was born in Chicago in 1971 and as a child and teenager became a keen and skilled photographer. He was impressed by the camera's ability to halt time by capturing a moment and was impressed by the work of Ansel Adams, Richard Avedon and photographs in *Life* magazine. Later he played college football, studied drama and worked as an actor. In 1994, while living in New York, he resumed photography and achieved success by shooting fashion and portraits of celebrities. Appalled by the 9/11 attacks on the Twin Towers, he visited the scene to take photographs but had to leave because of a severe asthmatic attack caused by the toxic chemical smoke. The Uniformed Firefighters Association and the Uniformed Fire Officers Association then appointed him the official archive photographer at Ground Zero. Using a Mamiya medium format camera for maximum image clarity, he documented the rescue and recovery efforts for six months and decided not to profit from any of his photographs by selling them. In 2002, a collection of his photographs was published in book form: *Requiem: Images of Ground Zero*.

Suson's sepia-toned portrait of a retired New York firefighter (ex-Rescue Company No. 2) with a sad, resigned facial expression represents an extremely poignant moment because the man – Lee Ielpi – is gazing down at a battered helmet held against his chest with his right arm. This helmet belonged to his deceased, eldest son Jonathan of Squad 288 North Queens who perished on 9/11. Although over 300 firefighters died, only eight helmets were found during the recovery operation. At the time of his death, Jonathan was aged 29, married with two children. When Lee heard his son was missing, he vowed to search Ground Zero until his body was found in order to 'bring him home', but it was not until 11 December that Jonathan's remains were discovered at the site of the South Tower. Lee has another son Brendan who is also a New York firefighter.

Suson's tender portrait has been reproduced on a

207 Gary Suson, *Portrait of Lee Ielpi*, (2001). Photograph reproduced on a postcard, 11 x 15 cms © Gary Suson, 2004

postcard published by maxracks.com (a national rack company that issues free postcards) to publicise septembereleven.net, a website he established 'in memory of all victims of terrorism'. At the time of writing, many of Suson's colour, black-and-white, and sepia-toned images can be viewed on this website.

## 208 — 2001

### Hearst Entertainment Inc., The Bravest Real Stories. Real Action. Real Heroes

In spring 2001, National Entertainment in association with Hearst Entertainment Inc., began filming an ambitious documentary or 'reality' series about firefighting for American and Canadian television. Later, video and DVD versions of extracts from the weekly hour-long episodes were released. Russell Best was the series creator and executive producer, Tammy Leech was co-executive producer and Lt. Peter M. Sapienza, an experienced New York firefighter, was its Technical Advisor. Altogether 22 episodes were made about fire departments across America, which included those in Brooklyn, Boston, Dallas, Los Angeles, Miami, Newark and Washington.

In an episode about San Francisco, the female firefighter Caroline Paul was featured. Paul (b. 1963), an athlete, journalist, aspiring filmmaker and Stanford University graduate, is the author of a fascinating and fluently written account of her career as a firefighter in a rescue unit from 1989 onwards entitled *Fighting Fire: A Personal Story*, (1998). Incidentally, her twin sister Alexandra, a star of *Baywatch*, played a firefighter in the television series *Fire Co. 132* (1997), which was aired internationally but not shown in the United States.

During the making of *The Bravest*, small cameras were installed in the front cabs, on the back steps and roofs of fire engines, and camera crews accompanied fire companies as they attended major and minor fires, medical emergencies and rescues, and dealt with hazardous chemicals. Intimate footage of firehouse life and interviews with firefighters stressed the comradeship, pride and enjoyment typical of their occupation. The officers talked frankly about their adrenaline rushes, families, fears and the dangers they faced.

In promoting the series, the makers vaunted its live-action realism and one of the means for achieving this was a new mini video camera called the 'Vest Cam' (invented by Mitchell Wagenberg, aka 'king of hidden cameras', of StreeTVision, Manhattan) that was attached to the bodies of firefighters before they went into action. The close-up sequences recorded inside buildings on fire with this camera conveyed a strong 'you are there' feeling.

Over 200 American television stations were scheduled to show *The Bravest* in the fall of 2001 but before it was aired, 9/11 occurred. Eighteen firefighters from Manhattan's Rescue 1 and Brooklyn's Rescue 2 – including Gary Geidel – who had been filmed in the summer, died on that day and so a new segment was introduced paying homage to the fallen. Naturally, public and media interest in the show was heightened by the 9/11 losses. Firefighters across the nation praised the series for representing their daily experiences from their point of view and with vivid accuracy.

208 Hearst Entertainment Inc., *The Bravest Real Stories. Real Action. Real Heroes*, (2001). Cover design of DVD based on a documentary television series, Hearst Entertainment, Inc

The DVD's wrapper photograph depicts the head and shoulders of a mature male firefighter who is wearing a helmet with protective earflaps. His lean features are illuminated from below by the glow of a fire. He has a grimy face, stubble on his chin and a determined but tired expression. It would seem he has been working hard all night. Since he resembles the film star Clint Eastwood, the portrait embodies a tough masculine ethos. Furthermore, the firefighter acquires a giant's stature by being placed in the sky above several other, smaller firefighters who appear as dark silhouettes against the yellow light of the blaze. Since they stand immobile holding the staffs of hooks, they resemble priests participating in a sacred ritual.

## 209 2001

**Alex Ross, Cover of Heroes: The World's Greatest Super Hero Creators Honor the World's Greatest Heroes 9–11–2001, Vol. 1, No. 1**

## 210 2002

**David Finch & Art Thibert, A scene from The Call of Duty: The Brotherhood, Vol. 1, No. 1**

Illustrators and writers working for American comic book publishers, most of which are located in New York, naturally reacted quickly to 9/11 and many of their images featured firefighters. The visual spectacle of the terrorist attack itself resembled scenes of urban destruction commonly found in comic books and Hollywood disaster movies. Comic book superheroes, of course, are fantasies that are impotent in the face of reality: after all, neither Superman nor Captain America prevented 9/11. Many wondered, therefore, how the comics industry would respond to the event and the fact that the new heroes of America were ordinary human beings employed as rescue workers. Journalists asked: 'Will 9/11 mark the end of superhero characters?'

209 Alex Ross (illustrator), Cover of Heroes: *The World's Greatest Super Hero Creators Honor the World's Greatest Heroes*, 9–11–2001, Vol. 1, No. 1, December 2001. Comic book edited by Joe Quesada. New York: Marvel Comics. © 2001 Marvel Characters. Author's collection

210 David Finch & Art Thibert (illustrators), A scene from *The Call of Duty: The Brotherhood*, Vol. 1, No. 1 August 2002. Comic book by Chuck Austin & Others. New York: Marvel Comics. © 2002 Marvel Characters. Author's collection

In an unprecedented spirit of collaboration, leading artists and writers created images and stories for free and comic book houses issued no less that seven publications about 9/11 to be sold for the benefit of charities. In terms of content, there were two main reactions: first, superheroes continued to be depicted but they were shocked and humbled by the attack on America: some grieved and shed tears, others were bruised and battered but also angry and defiant; second, the rescue workers themselves were represented as heroes who merited upgrading to superhero status.

Joe Quesada of Marvel Comics edited a compilation of 64 images in a variety of styles by 100 leading illustrators entitled *Heroes*, which included a few images of superheroes such as the Incredible Hulk and Captain America, but more images celebrating rescue workers. Certain illustrations depicted both superheroes and rescue workers. Arguably, superheroes play the same social role for many Americans as the mythical gods of antiquity did for the Greeks and Romans. In one illustration, the huge figures of Iron Man, Thor and Captain America holding extinguished candles occupied the background while rescue workers toiled in the foreground. Other drawings depicted rescue workers as giant, god-like figures. The message was: superheroes are restricted to the printed page but they can embody human emotions and, fortunately, America has real-life heroes. Alex Ross painted the book's sombre tonal cover, which showed a firefighter with his head bowed carrying the body of a victim away from the ruins of the Twin Towers.

Feelings and patriotism obviously ran high but too many illustrators relied on existing news photographs as sources and often the quality of the imagination and artwork was clichéd. (In a critique of comic books published in the wake of 9/11, R. Fiore of *The Comics Journal* argued that the event 'belonged to the camera and the camcorder. In order to represent something on paper an artist has to be able to comprehend it, and the destruction was incomprehensible.')

No artist depicted the forces responsible for the attack nor reflected on its causes. One of the best and most straightforward images, by Quesada, Todd McFarlane and Richard Isanove, showed an exhausted fireman covered in ash, with his head in his hands, recovering on a park bench. *Heroes* was published in December 2001 and sales soon approached a million. Proceeds were donated to the Twin Towers Fund and the original artwork was auctioned to benefit the same charity.

In January 2002, DC Comics Inc of New York, home of Batman and Superman, published a two-volume tribute book entitled 9/11, which contained contributions by a galaxy of comic book artists and writers. (Incidentally, DC Comics had previously issued comic books featuring firefighters, witness the character Fireman Fred Farrell who appeared in *Showcase* No.1, 1956.) In the same month, Alternative Comics of Gainesville, Florida published a similar book entitled *9–11: Emergency Relief*, a collection of first person, non-fiction accounts of the terrorist attack. Will Eisner, who has been credited with the invention of the graphic novel, was one of the contributors. New York's firefighters, of course, were represented. During January and February 2002, an exhibition of panels of artwork from this publication and, entitled 'Heroes Among Us', was mounted at the New York City Fire Museum in SoHo. Another exhibition took place at Exit Art in New York: this time the artwork displayed was from DC's 9/11 and *9/11: Artists Respond* published by the consortium Chaos, Dark Horse and Image Comics. Many comic book artists also donated original drawings (335 in total) to the Prints and Photographs Division of the Library of Congress.

Later, in August 2002, Marvel Comics began publication of *The Call of Duty*, a comic book series devoted to tales of courageous firefighters, police and other emergency aid workers, which Quesada has acknowledged was a response to the change of mood caused by 9/11. Chuck Austin, who is a television screenwriter and author of the first story about Lieutenant James MacDonald, a tough Brooklyn fireman, has described himself as a fire service buff. Illustrated by David Finch (pencil) and Art Thibert (inks), the story featured vivid scenes of firefighters battling a four-alarm blaze in a downtown brownstone. The images pulled no punches by showing a body on fire but the strip also included factual information about firefighting techniques and tools such as the Halligan (a forcible entry tool invented by New York fireman Huey Halligan), and the behaviour of fire trapped inside rooms. The appearance of a girl who seemed to be a ghost because she was immune from the flames, added a mystical dimension to an otherwise gritty, realistic narrative.

## 211 — 2002

**Jesus Romo, Fallen Brother**

## 212a — 2002

**Lawrence Noble, Holding the Line**

## 212b — 2002

**detail**

On 6 April 2002, the California Firefighters' Memorial in Capitol Park, adjoining the State Capitol in Sacramento, was unveiled in the presence of Governor Gray Davis, other dignitaries and several thousand uniformed firefighters and their families. The idea for the Memorial dated back to the early 1990s and was that of Dan Terry, President of the California Professional Firefighters, an organisation that represents 30,000 career firefighters and emergency medical personnel. A non-profit organisation was established – The California Fire Foundation – and by various means it raised the $2 million needed. A two-sided, three-part Memorial Wall (610 x 244 cms) made from polished limestone lists the names of 855 Californian firemen and women who have died on duty since 1850. Blank areas have been left for the addition of further names. At each end of the central wall are two smaller walls adorned with poignant bronze sculptures by Jesus Romo depicting a turnout coat and helmet suspended from a hook. Nearby are two freestanding sculptures of firefighters, one by Romo and one by Lawrence Alan Noble. Both sculptures are dramatic and represent moments of action. The setting is a quiet pleasant one of grass and trees.

Romo's bronze, which is the smaller of the two,

stands on a small circular base. It depicts a male firefighter carrying the limp bareheaded body of a dead comrade away from a fire ground. The face of the helmeted officer carrying the burden is anxious and strained. Romo has explained that he had wanted to portray 'the expression of desperation and pain and anguish that every firefighter feels when a brother goes down'. Essentially, therefore, the sculpture symbolises 'brotherhood'. Romo persuaded Gerry Shannon, a San Francisco Fire Captain, to serve as a model.

The 59–year-old Romo was born in Mexico City and came to the United States at the age of 18. He is not a professionally trained artist but a retired Battalion Chief of the Sacramento Fire Department. (Consequently, he had personal experience of losing colleagues.) However, he has been interested in art since childhood and for many years, has enjoyed painting and sculpting in his spare time. His sculpture is comparable in artistic quality

211 Jesus Romo, *Fallen Brother*, (2002). Memorial/public sculpture, bronze, 2.4 m high. Sacramento: California Firefighters' Memorial, Capitol Park. Photos courtesy of the California Professional Firefighters

212a Lawrence Noble, *Holding the Line*, (2002). Memorial/public sculpture, bronze and stone, 4 m high. Sacramento: California Firefighters' Memorial, Capitol Park. Photo courtesy of the artist

212b detail

to that of commercially successful American artists who design public monuments.

Fifty-three-year-old Noble, on the other hand, is an experienced artist and illustrator who runs a long-established studio – the Noble Studio – in Crestline, California, which specialises in the design and creation of large-scale, public sculptures. Noble, who was born in Tampa, Florida and educated in Houston, Texas, is extremely versatile: he has designed advertising campaigns for Hollywood movies, posters for presidential candidates, Olympic gold medals, a Star Wars chess set, bronze portraits of Civil War generals and famous entertainers, and a memorial to a police officer killed in the line of duty.

Noble's *Holding the Line* emphasizes the importance of teamwork in fighting fires: three figures feed a high pressure hose line forward to a fourth in charge of the nozzle; they advance on a gradually rising rocky base and so the helmet of the leading figure is over four metres above ground level. The figures are rather bulky because they are wearing heavy protective clothing and carry air tanks on their backs. Towards the rear, one firefighter reaches down to offer a helping hand to the last in line. Some viewers have found this gesture especially moving. Of course, the two hands nearly touching recalls Michelangelo's Sistine ceiling fresco of God creating mankind.

Noble wanted to signal the ethnic diversity of the Californian fire service and so he represented one figure as a Latino and another as an African-American. Unusually, one figure is female. Rose Conroy, a Fire Chief from Davis, California, modelled for this figure and felt honoured to represent the women firefighters of California. The face of another figure was derived from a photograph of Kenneth Earl Enslow who died in 1990 while fighting a forest fire. Noble modelled the figures and the Mussi Artworks Foundry & Gallery of Berkeley, California cast them.

## 213 — 2002

### Unknown photographer, A team of smokejumpers led by actor Wes Studi in a scene from Superfire

213 Unknown photographer, A team of smokejumpers led by actor Wes Studi in a scene from *Superfire*, (2002). Two-part, telemovie, 210 mins. Marina del Rey, CA: Lions Gate Television Inc., for ABC television. Photo courtesy of International Services, Lions Gate Entertainment Inc.

*Superfire* is a colour film, produced in Canada for American television, about fighting forest fires in Oregon, even though location shooting took place in Queensland, Australia and Tokoroa, New Zealand. It was first screened on ABC television in the United States in April 2002 and in Britain on Sky Movies Max in November 2002. Two brothers – Richard and Douglas Preston – who are best-selling authors wrote the script and Steven Quale – a first timer – directed it. The telemovie starred D.B. Sweeney as air tanker pilot James Merrick, Diane Farr as spotter plane pilot Sammy Kerns, and Wes Studi (a Cherokee Indian actor who often plays native Americans) as experienced smokejumper Joe Nighttrail.

Merrick arrives at the smokejumpers' base under a cloud of suspicion having made a mistake during a previous fire, which had caused a loss of life; consequently, he performs absurdly heroic deeds in order to redeem himself. Three small forest fires converge to create a so-called 'superfire' that threatens local communities and even the city of Portland. A team of smokejumpers, which includes one female, parachute into the forest and start to clear ground but soon they are trapped and all die except for one rookie even though they had wrapped themselves in fireproof sacks and Merrick had tried to drop water on them from his 50–year-old tanker aircraft. (We do not learn why the forest fire service has to rely on such antiquated equipment.) A superfluous sub-plot concerns a group of tedious teenagers stranded in the forest. 'Hotshots' arrive at the base but the film does not explain who they are and what they do. When the out-of-control conflagration threatens the lives and homes of local citizens, Merrick resorts to desperate measures: he drops a bomb from a low altitude on the fire to remove its oxygen. (Miraculously, his aircraft survives the blast.) By this means, he saves the day and is hailed a hero.

*Superfire* includes some vivid scenes of fire travelling at high speed and smokejumpers surrounded by walls of flame; hence, credit should be given to the special effects experts. It also conveys some basic information about how forest fires are fought but it is a mediocre production with a far-fetched plot. A thorough documentary such as *Fire* (1998), in the *Raging Planet* series, made by Pioneer Productions of London for television's Discovery Channel, is much more informative about the subject of fighting forest fires.

## 214 — 2002

### News Shopper staff photographer, Station Commander Danielle Cotton

In May 2002, local papers in south London carried a commonplace human-interest story about Danielle or

'Dany' Cotton, aged 32, who had been appointed Station Commander of the Eltham and East Greenwich fire stations. This made her the highest-ranking operational woman firefighter in Britain. During 2002, she was also named 'Outstanding Public Servant of the Year'. According to Fire Commissioner Brian Robertson, Cotton was 'a seasoned operational officer who has attended many serious incidents ... an articulate ambassador for the London Fire Brigade [who had made] a great contribution to furthering the cause of women in the fire service nationally'.

Cotton joined the London Fire Brigade at the age of 18 and besides the rigours of training and the job, she had to cope with bullying by some of her macho colleagues. She is still the only woman among the 56 male officers at the stations. (Only 103 of the capital's 5500 firefighters are women but a target of 15 per cent representation has been set for 2009.) Illustrating the articles were colour and monochrome photographs taken by staff photographers that depicted the fresh-faced Cotton in predictable ways: wearing protective clothing with a walkie-talkie radio and torch, and holding a helmet with a visor; sliding down a pole in the same outfit; and kneeling in front of a fire engine while holding a helmet in one hand and a hosepipe nozzle in the other. These images did not fully reflect her daily work because, as Cotton told reporters, as a manager she now had to spend much of her time in front of computer screens. Double entendres in headlines were predictable too; one read: 'Fire boss Dany is hot stuff.'

214 *News Shopper* staff photographer, *Station Commander Danielle Cotton*, (2002). Photograph published in *News Shopper*, 22 May 2002

## 215 2002

**Riddell, Fireman's Lift**

## 216 2003

**News Shopper staff photographer, Firefighters Jim Astley, Keith Jones, Pete Ives and Gerry Reeves keep warm while on strike**

In November 2002, Britain's Fire Brigades Union (FBU) – which has tens of thousands of members – called a national strike for only the second time in the 84 years of its existence. It was 25 years since their previous stoppage in 1977. The FBU was demanding a well-deserved 40 per cent pay increase in order to bring firefighters' wages of into line with other, similar professions and in fact, during negotiations, it was willing to settle for less. Many members of the public supported the strikers including the rock musician Joe Strummer (1952–2002, ex-leader of the punk band The Clash) of The Mescaleros. During a benefit gig by the band at Acton Town Hall, London on 15 November 2002, an FBU representative presented Strummer with a statuette of a firefighter by way of thanks.

However, part-time/volunteer firefighters belonging to the Retained Firefighters' Union (RFU, founded in 1976), operating in small towns and rural areas, refused to join the strike. Furthermore, right-wing newspapers carried headlines such as 'Flaming Greed' and 'Fire Union Bullies', and trotted out predictable complaints about the strikers: they were lazy and slept on the job; their shift system enabled them to do second jobs; they were manual workers not professionals; the fact that there were waiting lists to join the fire service must mean that it was already an adequately paid occupation; they were a group of mostly white males slow to accept women and members of the ethnic minorities; they engaged in restrictive practices and were unwilling to 'modernise' in order to increase efficiency and productivity (surely, the less they have to do the better?); by striking they were putting lives of citizens at risk. (Regarding the latter charge, firefighters left picket lines when lives were in danger and, of course, the lives of their families and friends were endangered, and their

215 Riddell, *Fireman's Lift*, (2002). Cartoon, pen and ink. *Observer*, 17 November 2002, p. 28. Reproduced courtesy of Chris Riddell

own lives were constantly at risk.) It would seem that when firefighters act as trade unionists and form picket lines, erstwhile heroes become – in the eyes of some – anti-heroes.

Naturally, the Labour Government under the leadership of Tony Blair, fearful of a wave of other public sector wage demands, was irritated and embarrassed by the strike and reluctant to provide any new money to pay the firefighters more. In Chris Riddell's cartoon, a huge grim-faced firefighter – personifying the FBU's members – grasps Blair by the genitals and lifts him in the air with one arm. (So, this cartoon is a variant on the motif of the fireman's lift discussed earlier.) A frightened and tearful Prime Minister cries out to 'Gordon' – that is, Gordon Brown, Chancellor of the Exchequer – for help. Clearly, Riddell believed the firefighters were more powerful than the politicians. However, the Government proved stubborn and the FBU had to settle in June 2003, after months of sporadic industrial action, for much less than they had wanted. Wage increases were also made conditional upon 'modernisation' of the fire service. Riddell's representation, therefore, was an over-optimistic one.

216 *News Shopper* staff photographer, *Firefighters Jim Astley, Keith Jones, Pete Ives and Gerry Reeves keep warm while on strike*, (2003). Photograph published in *News Shopper* (Lewisham edition), 22 January 2003

Riddell, a cartoonist, children's book illustrator and writer, was born in South Africa in 1962 but grew up in Britain. He attended Epsom School of Art & Design and studied illustration at Brighton Polytechnic's Art School. During the 1990s, he became political cartoonist for the Sunday broadsheet, the *Observer*. His drawing style has been described as continuing the symbolist tradition of *Punch* magazine. Riddell's cartoon, of course, was only one of many cartoons, photographs, film and video images depicting firefighters that appeared in the mass media during the course of the trade dispute. As the photograph published in *The News Shopper* demonstrates, the firemen-around-a-brazier motif was again popular with the press.

## 217 2002–03

### Robin Holland/ContentFilm, Sigourney Weaver and Anthony LaPaglia in a scene from the film The Guys

Theatre director Jim Simpson directed the low-budget American film *The Guys*, which was premiered at the Toronto International Film Festival on 11 September 2002 and released generally in the United States in March 2003. The film starred his wife Sigourney Weaver as 'Joan', an editor and writer, and the Australian-born actor Anthony LaPaglia as 'Nick', a NYFD Captain of Ladder Company No. 60. Simpson and Anne Nelson wrote its script.

Nelson, an ex-war reporter and Director of the International Program at Columbia University's Graduate School of Journalism, is the author of a one-act play with the same title that has been performed in the United States, Scotland and Ireland since December 2001. In 2002, Random House also published a book of the stage play, which had been penned in the immediate aftermath of 9/11. It was one of the first creative responses to the tragedy. In nine nights, Nelson wrote a series of monologues and dialogues (based on her personal experience) between a traumatised fire captain who had been assigned the onerous task of composing eulogies to eight of his men killed at the World Trade Center and a female journalist who had agreed to assist him 'find the words'. (The eulogies were to be delivered at memorial services.) Nelson will not reveal the name of the actual captain she helped because he requested anonymity.

217 Robin Holland/ContentFilm (photographer), Sigourney Weaver and Anthony LaPaglia in a scene from the film *The Guys*, (2002–03). Publicity photo © 2002 Focus Features. All Rights Reserved

As Joan questions and listens to Nick, she learns about the generosity, dependability and professionalism of firefighters, about the tasks they perform and the close-knit nature of firehouse culture, and the lives and different characters, skills, hobbies, etc., of the individuals who perished. The aim was to portray them as unique human beings rather than as superheroes. Naturally, Joan also finds out about Nick himself – the guilt he feels for being one of the survivors (he was off-duty at the time of the attack) and about his enthusiasm for dancing. She discovers that Nick's role as a manager includes pastoral care for which he was ill prepared. (So many died on 9/11 that the NYFD had to establish a special team to pen eulogies.) When the play was first performed off-Broadway, it attracted full houses and many firefighters and their families went to see it.

In the publicity photograph reproduced (a scene set in a church) and a poster for the film, LaPaglia wears a fire captain's smart dress uniform plus white cap and gloves. He stands motionless reverentially holding in both hands a firefighter's helmet. The latter, of course, serves as a synecdoche for 'the guys' who are now absent and whose verbal portraits he has to compose. In the play/film, the firefighters named were fictionalised composites of some of those who died. The film's climax occurs in a church in which Nick successfully delivers one of the eulogies Joan has written. Besides film extras, the congregation included members of families of the fallen firefighters.

No representation or political analysis of the terrorist attack is attempted because the film concentrates on the emotions of anger, grief, shock and sadness it caused. While the film is serious, it also has some humorous moments when jokes firefighters used to crack are recalled. Although some reviewers felt the film lacked drama and was superficial, 'mawkish testimony', others praised the strong performances of the two leading actors and described it as 'elegiac, a poignant tribute' and compared its social function to that of a war memorial or a shrine. Essentially, *The Guys* was a prime example of what Freud called, in his 1917 essay 'Mourning and melancholy', 'mourning-work', an emotional process that has to be gone through before healing can occur. The play/film also demonstrated the power of art to articulate feelings experienced by many millions and, via theatre audiences, to bring people together. As Simpson has remarked: 'Theater's obligation is to be more than escapist. It should offer an opportunity for the community to come together to encounter the catastrophe on human terms.'

218 Sergio Ostroverhy, *The Fireman*, (2003?). Pencil, paper and canvas, 2 x 2.5 m. Photo © 2004 Sergio Ostroverhy. All Rights Reserved

## 218 — 2003

## Sergio Ostroverhy, The Fireman

Ostroverhy (b. 1966) is a French prize-winning draughtsman, painter, art teacher, interior designer and translator based in Paris. Since he worked as a firefighter during his national military service (1988), he has first-hand knowledge of the occupation. Beginning in 1979, Ostroverhy undertook extensive training as a painter in academies and colleges in Budapest, Boston, Moscow and Paris. During the 1980s, he produced gouache and acrylic portraits of firemen, and Plaisance fire station in an expressionist manner heavily influenced by the work of Chaim Soutine.

His present style is photorealist in character – reminiscent of the work of the American Chuck Close – and he specialises in large-scale, close-up portraits either of individuals or groups of firemen (he never seems to paint any female firefighters). He has also painted portraits of Alpine hunters and French foreign legionnaires. When painting group portraits depicting whole figures, he pays particular attention to uniforms and equipment, and to the shiny surfaces of helmets. In 1997, Ostroverhy was awarded first prize at the New York City Museum's 'Firefighter's Annual Art Show' for his display of 300 portrait drawings of French firemen. He teaches master classes in traditional, academic, figurative techniques to students in a Montparnasse studio, directs his own company Fire Design Paris Ltd and maintains a website – www.fireface.com – to publicise his art and career.

Ostroverhy has described *The Fireman* as 'the world's largest fireman in painting' and has dedicated it to the firefighters of New York. Since the subject's bare head is motionless and his muscular features rather grim (the mouth turns down at the corners and the eyes gaze straight ahead into the distance), a strong sense of monumentality, masculinity and toughness is communicated. Ostroverhy thinks the human face conveys 'meaning' through 'a thousand expressions'; however, paradoxically, it is also 'a mask, a mystery, made of rock, the face of a mountain. Eternal ... the face of a hero'. It is, therefore, an appropriate image on which to conclude this book.

# Bibliography


A.F.S. Man (i.e. Vic Flint), *The Bells Go Down: The Diary of a London A.F.S. Man* (London: Methuen & Co., 1942).

Adrian, Colin, *Fighting Fire: A Century of Service 1884–1984* (Sydney, London, Boston: Allen & Unwin, 1984). History of the Sydney and New South Wales Fire Brigades.

Anderson, John James, *Our Fire-Fighters* (London: Oxford University Press, 1944).

Andersson, Kelly, 'The monument maker,' [Joyce Killebrew] *Wildland Firefighter*, Vol. 2, No. 3, July 1998, pp. 16–19.

Anon., 'The life of a New York Fireman,' *Harpers New Monthly Magazine*, Vol. LV, October 1877, pp. 659–72.

Archibald, Joseph, *The Smoke Eaters* (New York: David McKay, 1965).

Armstrong, Dee & others, *The Changing Face of the Fire Service: A Handbook on Women in Firefighting* (Madison, WI: Women in the Fire Service, 1993).

Arts Council of Great Britain, *Gustave Courbet 1819–1877* (London: Royal Academy of Arts, 1978).

Asbury, Herbert, *Ye Olde Fire Laddies* (New York: Alfred A. Knopf, 1930).

Auxiliary Fireman of the London Fire Brigade, *The Fire Fighters of London in Action* (Garden City, NY: Garden City Publishing Co., Inc., 1941), illustrated by members of the London Fire Service.

Backes, Nancy & Editors of Country Beautiful, *Great Fires of America* (Waukesha, WIS: Country Beautiful Corporation, 1973).

Baigent, Dave, *One More Last Working Class Hero: A Cultural Audit of the UK Fire Service* (Cambridge: Fitting-in Ltd & The Fire Service Research and Training Unit, Anglia Polytechnic University, 2001). Sociology PhD thesis: www.fitting-in.com.

Bailey, Victor (ed.), *Forged in Fire: The History of the Fire Brigades Union* (London: Lawrence & Wishart, 1992).

Baird, Donal M., *The Story of Firefighting in Canada* (Erin, Ontario: Boston Mills Press, c. 1986).

Baker, Alan R.H., '6: Fire-fighting Corps,' *Fraternity among the French Peasantry: Sociability and Voluntary Associations in the Loire Valley, 1815–1914* (Cambridge: Cambridge University Press, 1999), pp. 192–218.

Ballantyne, Robert Michael, *Fighting the Flames: A Tale of the London Fire Brigade* (London: James Nisbet & Co., 1868).

– *Life in the Red Brigade: A Story for Boys* (London: G. Routledge & Sons, 1873). *Life in the Red Brigade: A Fiery Tale bound with Fort Desolation or, Solitude in the Wilderness* (London: James Nisbet & Co., 1887).

Baumer, Edward, *The Early Days of the Sun Fire Office* (London: Sir Joseph Causton & Sons, 1910).

Bennett, Catherine, 'Every strike needs a brazier,' *Guardian*, G2, 28 November 2002, p. 5.

Bertin, François, Courault, Pascal & Deville, Joan, *La Grande Histoire des Sapeurs-Pompiers: Le Feu Sacré* (Paris: France Loisirs, 1994).

Betts, Ben, *Heroes with Grimy Faces ... Cartoons of Life in the Fire Service and the Lighter side of the Blitz* (London: C.A. Pearson, 1941), 32–page booklet of cartoons by an auxiliary fireman.

Bishop, Robert, *American Folk Sculpture* (New York: E.P. Dutton & Co., 1974).

Blackstone, Geoffrey Vaughan, *A History of the British Fire Service* (London: Routledge & Kegan Paul, 1957). 2nd edn, Fire Protection Association, 1996.

Boff, Chris, *Boys' Book of Fire-Fighting* (London: G. Routledge & Sons, 1939).

Bracewell, Michael, 'Happy Days,' *Guardian, Weekend*, 28 February 2004, pp. 28–31.

Bradbury, Ray, 'The fireman,' *Galaxy Science Fiction*, Vol. 1, No. 5, February 1951, pp. 4–61.

– *Fahrenheit 451* (New York: Ballantine Books, 1953). Illustrated by Joe Mugnaini.

Braidwood, James, *On the Construction of Fire-engines and Apparatus, the Training of Firemen, and the Method of Proceeding in Cases of Fire* (Edinburgh: Oliver & Boyd, 1830).

Brown, Roy Frederick & Thomson, William Stuart, *The Battle against Fire* (London: Abelard-Schuman, 1966).

Brumbaugh, Thomas B., 'The Kentucky Monuments of

Robert E. Launitz', *The Filson Club Historical Quarterly*, Vol. 43, April 1969, pp. 165–72.
Bulau, Alwin E., *Footprints of Assurance* (New York: Macmillan, 1953).
Burgess-Wise, David, *Fire Engines and Fire-Fighting* (London: Octopus, 1977).
Burridge, Anne, *Fireman's World* (Braunton: Merlin, 1983).
Butcher, Thomas Kennedy, *The Fire-fighters* (London: Batsford, 1977). Past into Present series.
Calder, Angus, *The Myth of the Blitz* (London: Jonathan Cape, 1991).
Cannon, Donald J. (ed.), *Heritage of Flames: The illustrated History of Early American Firefighting* (Garden City, NY: Doubleday, 1977).
Capasso, Nick, 'Vendome Memorial, Boston, Massachusetts,' *Sculpture*, Vol. 17, No. 6, July-August 1998, pp. 14–15.
Carp, Benjamin L., 'Fire of Liberty: firefighters, urban voluntary culture, and the Revolutionary Movement,' *The William & Mary Quarterly*, Vol. 58, No. 4, October 2001. See also:<http://www.historycooperative.org/journals/wm/58.4/carp.html>
Cart-Tanneur, Philippe, Viger, Raoul & Peltier, Claude, *Sapeurs Pompiers de Paris des Origines à nos Jours* (Evanston: Branding Iron Productions, 1985).
'Centurion' (Ralph Tuck) and Reginald Mills (illustrator), *"In the Service of the Nation": The N.F.S Goes into Action*, (London: National Fire Service Benevolent Fund/Raphael Tuck & Sons, 1944).
Chetkovich, Carol, *Real Heat: Gender and Race in the Urban Fire Service* (Piscataway, NJ: Rutgers University Press, 1997).
Childs, Charles & Lipke, William C., *Thomas Waterman Wood, PNA, 1823–1903* (Montpelier, Vermont: Wood Art Gallery, 1972).
Christie's, *The Royal Exchange Collection of Firemen's Badges, Related Paintings and Furniture* (London: Christie's South Kensington, 26 June 2001). Sale catalogue.
Coe, Andrew (ed.), *F.D.N.Y. An Illustrated History of the Fire Department of the City of New York, as seen through the Collection of the New York City Fire Museum* (New York: Odyssey Books & James Brown House, 2003).
Cohen, Stan, *A Pictorial History of Smokejumping* (Missoula, MT: Pictorial Histories Publishing Company, 1983). Has a chapter on the 1952 movie *Red Skies of Montana*.
Colbert, Elias & Chamberlin, Everett, *Chicago and the Great Conflagration* (Cincinnati & New York: C.F. Vent, 1872), with illustrations by Chapin and Gulick from photographic views taken at the spot. Abridged, reprinted edition published by Viking Press of New York in 1971.
Coleman, Chief Ronny J., 'Firefighters in the popular imagination', in *Firefighters*, ed. by JoEllen L. Kelly (Emmitsburg, MD: National Fallen Firefighters Foundation/Westport, CT: Hugh Lauter Levin Associates, Inc., 2003), pp. 112–35.
Collins, Donald, *Our Volunteer Firemen 1736–1982* (Ephrata, PA: Science Press, 1982).
Cooper, Robyn, 'Millais' "The Rescue": A Painting of a "Dreadful Interruption of Domestic Peace"' *Art History*, Vol. 9, No. 4, December 1986, pp. 471–86.
– 'The fireman: immaculate manhood,' *Journal of Popular Culture*, Vol. 28, No.4, spring 1995, pp. 139–70.
Costello, Augustine E., *Our Firemen: A History of the New York Fire Departments, Volunteer and Paid*, (New York: A.E. Costello, 1887). Reprinted by Knickerbocker Press of New York in 1997.
Cox, Ronald, *Oh, Captain Shaw: the Life Story of the First and Most Famous Chief of The London Fire Brigade* (London: Victor Green Publications, 1984).
Creighton, John, *Fire-fighting in Action: The Modern British Fire Service* (Poole: Blandford, 1985).
Croker, K.B., *Things Phoenix 1782–1982* (London: Phoenix Assurance, 1982).
Crump, Irving, *The Boys' Book of Firemen* (New York: Dodd, Mead & Company, 1916).
– *The Boy Scout Fire Fighters* (New York: Barse & Hopkins, 1917).
– *Our Firemen* (New York: Dodd, Mead & Company, 1938).
Curtiss, Richard H., 'Giant screen film records horrors of Kuwait's flaming oil,' *Washington Report*, (March 1995), pp. 30, 106.
Dabakis, Melissa, *Visualizing Labor in American Sculpture: Monuments, Manliness, and the Work Ethic, 1880–1935* (New York: Cambridge University Press, 1999).
Dana, David, *The Fireman: The Fire Departments of the United States, with a Full Account of All Large Fires, Statistics of Losses ...* (Boston, MA: J. French & Co., 1858). Reprinted by Brohan Press in 1999.
Davies, E.A., *An Account of the Formation and Early Years of the Westminster Fire Office* (London: Country Life Ltd/Westminster Fire Office, 1952).
De la Plante, Allan (writer & photographer), *Flame of Courage: The Canadian Firefighter* (Etibocoke, ON: Window Publishing, 1993).
Deluca, Chuck, *Firehouse Memorabilia: A Collector's Reference Book* (York, ME: Maritime Antique Auctions, 1989).
Demarne, Cyril, *The London Blitz: A Fireman's Tale* (London: Parents' Centre Publications, Newham Parents' Centre, 1980).
– *Our Girls: A Story of the Nation's Wartime Firewomen* (Bishop Auckland, Durham: Pentland Press, 1995).
– 'Fifty years on. The story of the Firemen's Memorial,' *In Attendance: The Magazine of the British Firefighter*, Vol. 5, No. 10, September 1991, pp. 26–28.
Deroo, Raymond, Havard, Patrice & Ranvoisy, Emmanuel, *La Fabuleuse Histoire des Pompiers* (Paris: Tallandier Éditions, 2002).
Diekmann, Stefanie, 'Hero and superhero,' *Guardian*, Review, 24 April 2004, p. 19.
Ditzel, Paul C., *Fire Engines, Firefighters: The Men, Equipment and Machines, from Colonial Days to the Present* (New York: Crown Publishers, 1976).
– *A Century of Service, 1886–1986: The Centennial History of the Los Angeles Fire Department* (Los Angeles, CA: Los Angeles Firemen's Relief Association, c. 1986).
– *Firefighting during World War II* (New Albany, IND: Fire Buff House Publishers, c. 1994).
Dudley, Ernest, *Chance and the Fire-Horses: A Story that Really*

*Happened ...* (London: Harvill Press, 1972).
Dunshee, Kenneth Holcomb, *Enjine!- Enjine! A Story of Fire Protection* (New York: Harold Vincent Smith of The Home Insurance Company, 1939). Reprinted by Manor Publishing in 1980. Has a chapter on the H.V. Smith Museum Collection.
– *As You Pass By* (New York: Hastings House, 1952).
Earnest, Ernest, *The Volunteer Fire Company: Past and Present* (New York: Stein & Day, 1979).
Editors of *Vanity Fair*, 'One Week in September', *Vanity Fair*, November 2001. Special edition.
Engle, Marsh & Gay Reboli (photographer), *Amazing Women, Amazing Firefighters* (San Diego, CA: Jodere Group, 2002).
Ernest Brown & Phillips Ltd., *Catalogue of an Exhibition of Paintings and Drawings of the "Firefighters of London" by Auxiliary Fireman [Rudolf] Haybrook* (London: The Leicester Galleries, 1940).
Essen, Thomas von (ed.), *In the Line of Duty: A Tribute to New York's Finest and Bravest* (New York: HarperCollins, 2001).
– *Strong at Heart: Life and Death in the Fire Department of New York* (New York: ReganBooks, 2002).
*Exhibition of Paintings by Firemen Artists* (London: Royal Academy, 1941).
Fairbanks, Eugene F., *A Sculptor's Testimony in Bronze and Stone: Sacred Sculpture of Avard T. Fairbanks* (Salt Lake City, UT: Publisher's Press, c. 1972; E.F. Fairbanks, rev. edn, c. 1994).
Falk, Bernard, *Thomas Rowlandson: His Life and Art* (London: Hutchinson, 1949).
*Fallen Heroes: A Tribute from Fire Engineering, September 11, 2001* (Tulsa, OK: PennWell Books, 2001).
Farningham, Elizabeth, *George Halse, Sculptor, 1826–1895* (Doncaster: L.E. Farningham, 2002).
Felton, Harold W. & Watson, Aldren A. (illustrator), *Fire Fightin Mose: Being an Account of the Life and Times of the World's Greatest Fire Fighter, Member of the NYC Volunteer Fire Dept. of the Company of Lady Washington, Engine No. 40, a Machine of Greatest Excellence known as the White Ghost* (New York: Knopf, 1955).
– & Laite, Gordon (illustrator), *Big Mose, Hero Fireman* (Champaign, IL: Garrard Publishing Co., 1969).
Fiore, R. 'A moment of noise,' *The Comics Journal*, No. 247, October 2002.
Fire Brigades Union, *Twenty Years Ago we realised things would never be the same again. Reflections of the 1977 Strike* (London: Co-operative Press, 1997?).
Firebrace, Aylmer, Sir, *Fire Service Memories* (London & New York: A. Melrose, 1949).
*Firefighters: 30 Postcards* (Norwich: Jarrold Publishing, 1998).
Firemen Artists Organising Committee of the National Fire Service, *Exhibition of Paintings, Watercolours and Drawings by Firemen Artists* (London: Royal Exchange, 1943). Folding sheet.
Firemen's Artists Committee, *Exhibition of Paintings by Firemen Artists*, (London: Royal Academy, 1944).
Floherty, John J., *Fire Alarm: The Story of Fire Fighting* (Philadelphia & New York: Lippincott Company, 1949).
Flynn, Sean, *3000 Degrees: The True Story of a Deadly Fire and the Men who Fought It* (New York: Warner Books, 2002).
Ford, James L., *The Third Alarm: A Story of the New York Fire Department* (New York: Brentano's, 1893).
Foster, Ted, *Bushfire: History, Prevention, Control* (Sydney: Reed, 1976).
Fothergill, George Algernon, *British Fire-marks from 1680* (Edinburgh & London: W. Green & Sons, 1911).
Fowler, Miriam Rogers, *Giuseppe Moretti: Master Sculptor and Father of Vulcan* (Birmingham, AL: Birmingham Museum of Art, 2002).
Frank, Paul Arthur, *Das Deutsche Feuerwehrbuch* (Dresden und Wien: Dresdner Verlagsbuchhandlung Max Otto Groh, 1929).
Freedman, Jill (photographer) & Smith, Dennis (text), *Firehouse* (Garden City, NY: Doubleday, 1977).
Gale Research Company, *Currier & Ives: A Catalogue Raisonné* (Detroit, MI: Gale Research, c. 1984).
Garvan, Anthony N.B. & Wojtowicz, Carol A., *Catalogue of the Green Tree Collection* (Philadelphia, PA: Mutual Assurance Co., 1977).
Gauroy, Gerard, *Le Brigade de Paris* (Paris: Éditions Fernand Lanore, 1981).
Gibbs, Gladys T. (compiler), *Fighting Fire, with forty Pictures telling the Story of the Brave and Fearless Lives of Firemen* (Racine, WI: Whitman Publishing Company, 1934).
Golway, Terry, *So Others Might Live: A History of New York's Bravest, the FDNY from 1700 to the Present* (New York: Basic Books, 2002).
Goodenough, Simon, *Fire! The Story of the Fire Engine* (London: Orbis Publishing, 1978).
Goodman, Edward C., *Fire! The 100 Most Devastating Fires Through the Ages & the Heroes who Fought Them* (New York: Black Dog & Leventhal Publishers Inc., 2002).
Gorrell, Gena K., *Catching Fire: The Story of Firefighting* (Plattsburgh, NY: Tundra Books of Northern New York, c. 1999).
Gottschalk, Jack, *Firefighting* (New York: DK Publishing, Inc., 2002).
Grafton, Carol Belanger, *Trades and Occupations: A Pictorial Archive from Early Sources* (New York: Dover, 1990).
Granito, John A., 'Origins of firefighting in the United States', in *Firefighters*, ed. by JoEllen L. Kelly (Emmitsburg, MD: National Fallen Firefighters Foundation/Westport, CT: Hugh Lauter Levin Associates, Inc., 2003), pp. 14–35.
Gray, Frank, 'James Williamson's rescue narratives,' *Young and Innocent? The Cinema in Britain, 1896–1930*, ed. Andrew Higson (Exeter: University of Exeter Press, 2002).
Green, Floride, *Some Personal Recollections of Lillie Hitchcock Coit–5* (San Francisco, CA: Grabhorn, 1935).
Green-Hughes, Evan, *A History of Firefighting* (Ashbourne, Derbyshire: Moorland Publishing, 1979).
Greenberg, Amy S., *Cause for Alarm: The Volunteer Fire Department in the Nineteenth-Century City* (Princeton, NJ: Princeton University Press, 1988).
Greene, Douglas G., & Hearn, Michael Patrick, *W.W. Denslow*, (Mount Pleasant, MI: Clarke Historical Library/Central Michigan University, 1976).
Griffin, Alan, *This Noble Duty: A History of Firefighting in Warwickshire* (Warwickshire: Feldon, 1989).
Grissom, Carol A., 'Saving firemen made of zinc',

(Washington, DC: Smithsonian Center for Materials Research and Education, 2002). <www.si.edu/scmre/learning/zincefiremen.htm>
Groce, George C. & Wallace, David H., *New-York Historical Society Dictionary of Artists in America, 1564–1860* (New Haven, CT: Yale University Press, 1957).
Gulliford, Andrew, 'Fire on the mountain: tragic death and memorialization of the Storm King Fourteen,' *Montana: The Magazine of Western History*, summer 1997, pp. 44–57.
Gurka, Andrew G., *Hot Stuff: Firefighting Collectibles* (Marion, IN: LW Publishing & Book Sales, 2nd edn 2001).
Hagen, Susan & Carouba, Mary, (compilers), *Women at Ground Zero: Stories of Courage and Compassion* (Indianapolis, IN: Alpha Books, c. 2002).
Halberstam, David, *Firehouse* (New York: Hyperion, 2002).
Hall, George (photographer) & John Burks (writer), *Working Fire: The San Francisco Fire Department* (San Francisco, CA: Chronicle Books, 1985).
Hallahan, Pat, *"When the Bells Go Down,"* (Ilford, Essex: Pat Hallahan, 1983). Book of cartoons.
Hamilton, Richard R. & Charles N. Barnard, *20,000 Alarms: The Memoirs of New York's Most Decorated Fireman* (Chicago, IL: Playboy Press, 1975).
Hansen, Charles V., *The History of American Firefighting Toys: The Development of Real Equipment and a Collector's Guide to the Toys they Inspired* (Sykesville, MD: Greenberg Publishing Co., c. 1990).
Hansen, Jon, *Oklahoma Rescue: The Heroic Untold Story – Direct from the Heartland and the Heart* (New York: Ballantine Books, 1995).
Hashagen, Paul & Kimmerly, Janet, *Fire Department City of New York: The Bravest, an illustrated History 1865 to 2002* (Paducah, KY: Turner Publishing, 2002).
Hass, Ed., *Fire Equipment* (San Diego, CA: Thunder Bay Press, 1998).
Haydon, Arthur Lincoln, *The Book of the Fire Brigade: A Record of Fire-fighting from the Earliest Times to the Present Day ...* (London: Pilgrim Press, 1913).
Hayward, Charles F., *General Alarm: A Dramatic Account of Fires and Fire-Fighting in America* (New York: Dodd, Mead & Co., 1967).
Hazen, Margaret Hindle & Robert M., *Keepers of the Flame: The Role of Fire in American Culture 1775–1925* (Princeton, NJ: Princeton University Press, 1992).
Hendrickson, Noel (photographer) & Greer, David (text), *Coming through Fire: The Wildland Firefighter Experience* (Vancouver, Canada: Raincoast Books, 2001).
Henham, Brian, *True Hero: the Life and Times of James Braidwood, Father of the British Fire Service* (Romford: Braidwood Books, 2000).
– & Sharp, Brian, *Badges of Extinction: The 18th and 19th Century Badges of Insurance Office Firemen* (London: Quiller Press, 1989).
Heyden (or Heiden), Jan van der (elder and younger), *A Description of Fire Engines with Water Hoses and the Method of Fighting Fires now used in Amsterdam* (Canton, MA: Science History Publications, 1996). Translation by Lettie Stibbe Multhauf of 2nd edn of book published in 1735. 1st edn published in 1690.
Hill, Charles T., *Fighting a Fire* (New York: The Century Co., 1897). Illustrated with drawings by the author.
Hill, Draper & Muggeridge, Malcolm, *Illingworth on Target* (Boston, MA: Wiggin Gallery of Boston Public Library, 1970).
Holdridge, Helen, *Firebelle Lillie* (New York: Meredith, 1967).
Holloway, Sally, *London's Noble Fire Brigades 1833–1904* (London: Cassell, 1973).
– *Courage High! A History of Firefighting in London* (London: HSMO, 1992).
Holmes, Frederic Morrell, *Firemen and their Exploits: with Some Accounts of the Rise and Development of Fire-Brigades, etc.* (London: S.W. Partridge & Co., 1899).
Holzman, Robert S., *The Romance of Firefighting* (New York: Harper & Bros., 1956).
Home Insurance Company, *Firefighting Museum of The Home Insurance Company* (New York: Home Insurance Company, n.d.), 16-page pictorial guide.
Honeycombe, Gordon, *Red Watch: A True Story of a Fatal Fire and the Men who Fought It* (London: Hutchinson, 1976).
Hopkinson, Tom (ed.), *Picture Post 1938–50* (London: Allen Lane, 1970).
Hornung-Arnegg, Wolfgang, *Feuerwehrgeschichte: Brandschutz und Löschgerätetechnik von der Antike bis zur Gegenwart* (Stuttgart: Kohlhammer, 1955).
Hunter, Kenny & others, *Citizen Firefighter* (Glasgow: Centre for Contemporary Arts & Strathclyde Fire Brigade, 2002).
Hurst, Randle M., *The Smokejumpers* (Caldwell, ID: Caxton Printers, 1966).
Ingham, H.S. (ed.), *Fire and Water: An Anthology by Members of the NFS* (London: Lindsay Drummond Ltd., 1942). With drawings by firemen artists. Reprint: Exeter: Firestorm Publications, 1992.
Ingram, Arthur, *A History of Fire-Fighting and Equipment* (London: New English Library, 1978).
Insurance Company of North America, *American Fire Marks, the Insurance Company of North America Collection ...* (Philadelphia, PA: Insurance Company of North America, 1933).
Irvine, Louise & Atterbury, Paul [& others], *Gilbert Bayes: Sculptor 1872–1953* (Shepton Beauchamp, Somerset: Richard Dennis, 1998).
Israel, Barbara, 'The metalwork of J.W. Fiske and Company', *Magazine Antiques*, March 2000.
Jackson, W. Eric, *London's Fire Brigades* (London: Longmans, Green & Co., 1966).
Jewett, Masha Zakheim, *Coit Tower, San Francisco: Its History and Art* (Volcano, CA: Volcano Press, 1983).
Johnson, David, *The City Ablaze: The Second Great Fire of London 29th December 1940*, (London: William Kimber, 1980).
Jopson, Debra, *The Red Brigade: The Official History of the New South Wales Fire Brigade Employee's Union* (Sydney: NSW Fire Brigade's Employee's Union, 1994).
Jorgensen, Nels, *Smoke Jumpers* (New York: Bouregy & Curl, Inc., 1954), illustrated by Carl Kidwell. British edition: London & Glasgow: The Children's Press, n.d.
Kee, Robert (ed.), *The Picture Post Album* (London: Barrie &

Jenkins, 1989).
Kelly, JoEllen L. (ed. in chief), *Firefighters* (Emmitsburg, MD: National Fallen Firefighters Foundation/Westport, CT: Hugh Lauter Levin Associates, Inc., 2003).
Kenlon, John, *Fires and Fire-fighters: A History of Modern Fire-fighting with a Review of its Development from Earliest Times* (New York: George H. Doran, 1913).
Kenyon, James W., *The Fourth Arm: A Survey of Fire-Fighting, Past, Present and Future* (London: Harrap, 1948).
Kirk, John L. (ed.), *History of Fire Fighting* (York: Castle Museum, 2nd edn 1960).
Kerik, Bernard B. & Essen, Thomas von (forewords), *In the Line of Duty: A Tribute to New York's Bravest and Finest* (New York: Regan Books/HarperCollins, 2001).
Kelly, William W., 'Incendiary actions: fires and firefighting in the Shogun's capital and the People's City,' in Edo and Paris: *Urban Life and the State in the Early Modern Era*, eds James L. McClain & others (Ithaca & London: Cornell University Press, 1994), pp. 310–31.
Kinley, Jessie D., *Call Kinley: Adventures of an Oil Well Firefighter* (Tulsa, OK: Cock-a-Hoop Publishing, 1995).
Kitton, Frederic George, *Dickens and His Illustrators: Cruikshank ...* (London: George Redway, 1899).
Kogan, Herman & Cromie, Robert, *The Great Fire: Chicago* 1871 (New York: Putnam, 1971).
Kútvölgyi, Mihály, *Hungarian Fireguard Mementoes* (Budapest: TIMP, 2001).
Langewiesche, William, *American Ground: Unbuilding the World Trade Center* (New York: Farrar, Straus & Giroux, 2002).
Leschak, Peter M., *Ghosts of the Fireground: Echoes of the Great Peshtigo Fire and the Calling of a Wildland Firefighter* (San Francisco, CA: HarperSanFrancisco, 2002).
Lewis, David Andrew, *Didja' Ever Want To Be a FIREMAN?* (Oneonta, NY: State University of New York College at Oneonta, 2002). Thesis published via the Internet: www. auroraregionalfiremuseum.org/dija/didja.pdf.
*Life* magazine, 'London Firefighter: the story of one of Britain's new heroes,' *Life* magazine, 21 October 1940. Illustrated article about AFS Section Officer Louis 'Tug' Wilson and his crew of six.
– Robert Sullivan (ed.), *One Nation: America Remembers September* 11, 2001 (Boston, New York & London: Little, Brown & Company, 2001).
– & McNally, Joe (photographer), *Faces of Ground Zero: Portraits of the Heroes of September 11, 2001 by Joe McNally* (Boston, MA: Little, Brown, 2002).
Lloyd-Elliott, Martin, *City Ablaze: Life with the World's Busiest Fire-fighters* (London: Bloomsbury, 1992).
Loeper, John J., *By Hook & Ladder: The Story of Fire Fighting in America* (New York: Atheneum, 1981).
Lombardo, Josef Vincent, *Attilio Piccirilli: Life of an American Sculptor* (New York & Chicago: Pitman Publishing, 1945).
*London's Hour, As seen through the Eyes of the Fire-Fighters* (London: Staples Books Ltd, 1942).
Luke, R.H. & McArthur, A.G., *Bushfires in Australia* (Canberra: Australian Government Publishing, 1978).
Lussier, Hubert, *Les Sapeurs-Pompiers au XIXe Siècle, Associations Volontaires en Milieu* (Paris: A.R.F. Éditions/L'Harmattan, 1988).
Lyons, Paul Robert, *Fire in America* (Boston, MA: National Fire Protection Association, 1976).
McCole, John, *The Second Tower's Down: A Firefighter's Story* (London: Robson Books, 2002).
McCourt, Frank & others, *Brotherhood: In Strength and Sorrow – Images of the FDNY* (New York: Ogilvy & Mather/American Express Publishing, 2001).
McKay, Gary, *Firefighters: The Men and Women who Risk Their Lives to Save Ours* (Crows Nest, New South Wales: Allen & Unwin, 2002).
McLean, Gavin, *New Zealand Tragedies: Fires & Firefighting* (Wellington, New Zealand: Grantham House, 1992).
Maclean, John N., *Fire on the Mountain: The True Story of the South Canyon Fire* (New York: William Morrow & Co., 1999).
Magnum Photographers, *New York: September 11* (New York: Powerhouse Books, 2001)
Maidment, Brian, *Reading Popular Prints, 1790–1870*, (Manchester & New York: Manchester University Press, 2nd edn, 2001).
Man, Felix H., *Reportage Portraits* 1929–76 (London: National Portrait Gallery, 1976).
– *Man with Camera: Photographs from Seven Decades by Felix H. Man* (London: Secker & Warburg/New York: Schocken Books, 1983)
– *60 Years of Photography* (London: Victoria & Albert Museum, 1983)
Masters, Robert V. & Heavey, Ed (photographer), *Pictorial History of Fire Fighting: Going to Blazes* (New York: Castle Books/Sterling Publishing Co., 1950).
Micheels, Peter, *Braving the Flames* (New York: Berkley Publishing Group, 1989).
Ministry of Information for Ministry of Home Security, *Front Line 1940–41: The Official Story of the Civil Defence of Britain* (London: HMSO, 1942).
*Modern Boy's Book of Fire Fighters* (London: Amalgamated Press, 1939).
Moore, Charles (photographer), Durham, Michael S., & Young, Andrew (texts), *Powerful Days: The Civil Rights Photography of Charles Moore* (New York: Stewart, Tabori & Chang/Eastman Kodak Company, 1991).
Morris, Cyril Clarke Boville, *Fire!* (London & Glasgow: Blackie & Son Ltd, 1939).
Morris, John V., *Fires and Firefighting* (New York: Bramhall House, 1953).
Moses, Frederick T., *Firemen of Industry: 1854–1954, The Hundredth Anniversary of Firemen's Mutual Insurance Company* (Providence, RI: Firemen's Mutual Insurance Company, 1954).
Mundell, Frank, *Stories of the Fire Brigade* (London: Sunday School Union, 1895).
Murphy, Jim, *The Great Fire [Chicago, 1871]* (New York: Scholastic Inc., 1995).
Nelson, Anne, *The Guys: A Play* (New York: Random House, 2002).
Nicholls, Arthur, *Going to Blazes* (London: Hale, 1978).
Nissen, Beth, 'Firefighters: America's real-life superheroes,'<www.cnn.com/2001/US/11/01/rec.firefighter.heroes>

Nochlin, Linda, *Realism* (Harmondsworth, Middlesex: Penguin Books, 1971).
Novak, Michael J., 'Firemen in early photographs,' *The Photographic Historian*, Vol 6. No. 3, autumn 1985, pp. 8–14, plus front cover.
– 'The American volunteer Fireman and the Daguerreotype,' *The Daguerreian Annual* 1997 (Pittsburgh, PA: The Daguerreian Society, 1998), pp. 88–137.
'Oklahoma City, April 19, 1995,' *Newsweek*, May 1st 1995.
Osborne, Mary Pope (writer) & Johnson, Steve & Fancher, Lou (illustrators), *New York's Bravest* (New York: Alfred A. Knopf, 2002).
Otis, James, *An Amateur Fireman* (New York: E.P. Dutton & Co., 1898). Illustrated boy's fiction.
Page, Michael & Bryant, Malcolm, *Muscle and Pluck Forever: The South Australian Fire Service, 1840–1982* (Adelaide: South Australian Metropolitan Fire Services, 1983).
Patten, Robert L., *George Cruikshank's Life, Times and Art*, Vol. 1 1792–1835 (London: Lutterworth, 1992).
Paul, Caroline, *Fighting Fire: A Personal Story* (New York: St Martin's Press, 1998).
Paul, Tony, *The Story of the Fire Service* (London: Almark Publishing Co. Ltd., 1975).
Pendergast, Anthony W. & Ware, W. Porter, *Cigar Store Figures in American Folk Art* (Chicago, IL: Lightner Pub. Corp., 1953).
Petretti, Stephen, *Hotshots: F.N.G.* (San Marcos, AA: Flying Dutchman, 2002).
Phelps, Donald, 'Holman's legacy to popular humor: the disarranged landscape of Smokey Stover,' *The Comics Journal*, Vol 1, special edition, winter 2002, pp. 23–31.
Philadelphia Fire Department Historical Corporation, *Hike Out: The History of the Philadelphia Fire Department* (Philadelphia, PA: Philadelphia Fire Department Historical Corporation, 1999).
Phillips, *Firemarks and Other Relics Illustrating the History of Fire Insurance* (London: Phillips, 10 April 1979). Sale catalogue.
Piatti, James G. & Piatti, Sandra Frost, *Firefighting Antiques & Memorabilia* (Atglen, PA: Schiffer Publishing Ltd., 2001).
Picciotto, Richard 'Pitch' & Paisner, Daniel, *Last Man Down: A Firefighter's Story of Survival and Escape from the World Trade Center* (New York: Berkley Books, 2002).
Pickering, Harold 'Poony', *Poonytoons: The 'Lighter Side' of the Fire Service* (Stillwater, OK: Fire Protection Publications, Oklahoma State University, 1989).
Pickett, George, *The Brave: A Story of New York City's Firefighters* (New York: Brick Towers Press, 2002).
*The Picture Book of Firemen* (London & Melbourne: Ward, Lock & Co., c. 1935).
Porter, Stephen, 'Town fires: the case of Tiverton,' *Devon & Cornwall Notes & Queries* Vol. 33, No. 9, 1977, pp. 345–48.
Press Association, *9/11: A Tribute* (Rochester, Kent: Grange Books, 2003).
Pyne, Stephen J., *Fire in America: A Cultural History of Wildland and Rural Fire* (Princeton, NJ: Princeton University Press, 1982).
– *Burning Bush: A Fire History of Australia* (New York: Henry Holt & Company, 1991).
– *Year of the Fires: The Story of the Great [forest] Fires of 1910* (New York: Viking Press, 2001).
Rawls, Walton H., *The Great Book of Currier & Ives' America* (New York: Abbeville Press, c. 1979).
Reid, Alexander, *Aye Ready! The History of Edinburgh Fire Brigade, the Oldest Municipal Brigade in Britain* (Edinburgh: South-Eastern Fire Brigade/Geo Stewart & Co., 1974).
Reid, John Cowie, *Bucks and Bruisers: Pierce Egan and Regency England* (London: Routledge & Kegan Paul, 1971).
Reynolds, Donald Martin, *Monuments and Masterpieces: Histories and Views of Public Sculpture in New York City* (New York: Thames & Hudson, 2nd edn 1997).
Reynolds, P.K. Baillie, *The Vigiles of Imperial Rome* (London: Oxford University Press, 1926). Reprinted by Ares Publishers, Chicago in 1996.
Richards, Jeffrey, 'Fires were Started,' *History Today*, Vol. 45, No. 4 April 1995, pp. 29–34.
Riis, Jacob A., 'Heroes who fight fire' *The Century Magazine* Vol. 55, No. 4, February 1898, pp. 483–97.
Roetter, Charles, *Fire is Their Enemy* (London: Angus & Robertson, 1962).
Rogers, Donna, 'Images of heroes,' *Photographics Processing Magazine*, November 2001. On the work of the NYFD Photo Unit.
Ross, Alan, *Colours of War, War Art 1939–45*, (London: Jonathan Cape, 1983).
Rubin, James H., *Courbet* (London: Phaidon Press Ltd., 1997).
Rutledge, Anna Wells, *Catalogue of Paintings and Sculptures in the Council Chamber, City Hall, Charleston, South Carolina* (Charleston, SC: The City Council of Charleston, 1943).
Ryan, Keith, *The Illustrated History of Fire Engines* (Royston, Herts: Eagle Editions, 2002).
Sablayrolles, Robert, *Libertinus Miles: Les Cohertes Vigiles* (Rome: École Française de Rome, 1996).
Sachs, Edwin O., *A Record of the International Fire Exhibition, Earl's Court, London 1903 organised under the Auspices of the British Fire Prevention Committee* (London: British Fire Prevention Committee, 1903).
[Sansom, William], *Fire over London: The Story of the London Fire Service, 1940–41*, (London & Melbourne: London County Council/Hutchinson, 1941).
– Gordon, James & Spender, Stephen, *Jim Braidy: The Story of Britain's Firemen* (London: Lindsay Drummond, 1943). Illustrated by the work of firemen artists.
Sargent, Viktoria. '"Cool guys who are good guys": cultural representations of firefighters as heroes,' (Nottingham University, 2002) <www.fitting-in.com/sargent.htm>
Scher, Steven, *New York City Firefighting, 1901–2001* (Charleston, SC: Arcadia Publishing, 2002).
Seigel, Jessica, 'Firefighter chic: blue-collar reality and the folk hero archetype,' *The Village Voice*, 31 October – November 6, 2001.
Selway, Neville Carr, *The Regency Road: The Coaching Prints of James Pollard ...* (London: Faber & Faber, 1957).
– *James Pollard, 1792–1867: Painter of the Age of Coaching* (Leigh-on-Sea: F. Lewis, 1965).
Shaffer, Dale E., 'The sculptor of Salem: John G. Segesman,'

*Timeline* [Ohio Historical Society], Vol. 16, No. 1, January-February 1999, pp. 32–41.
Shapiro, Larry, *Firefighters* (St Paul, MN.: MBI (Motor Books International) Pub Co., 2003).
Shaw, Eyre Massey, Sir, *Fire Protection: A Complete Manual of the Organisation, Machinery, Discipline, and General Working of the Fire Brigade of London* (London: Charles & Edward Layton, 1876). A new, revised edition was published in 1890.
– *The Business of a Fireman: A Selection from the Writings of Sir E.M. Shaw* [made by Frank Eyre] (London: Lomax & Erskine & Co., 1945).
Sheldon, George William, *The Story of the Volunteer Fire Department of the City of New York* ... (New York: Harper, 1882).
Shippen, Katherine Binney & Graham, Margaret Bloy (illustrator), *Big Mose* (New York: Harper, 1953).
Singerman, Philip, *Red Adair: An American Hero, An Authorized Biography* (London: Bloomsbury, 1989).
Smith, Barry D., *Rescuers in Action: On Scene with America's Rescue Teams* (St Louis, MO: Mosby Lifeline, c. 1996).
Smith, Dennis, *Report from Engine Co. 82* (New York: McCall Books, 1972).
– *Dennis Smith's History of Firefighting in America: 300 Years of Courage* (New York: Dial Press, 1978).
– *Firefighters: Their Lives in Their Own Words* (New York: Doubleday, 1988).
– [and others], *Report from Ground Zero: The Story of the Rescue Efforts at the World Trade Center* (New York: Viking Press, 2002). Paperback edition published in UK by Corgi Books under the title *Report from Ground Zero: The Heroic Story of the Rescuers at the World Trade Center* (2002).
Smith, Elmer L. (ed.), *Fire Fighting at the Turn of the Century* (Lebanon, PA: Applied Arts Publishing, 1971).
Smith, Gary N., 'Mat Hastings: artistic chronicler of his times,' *Gateway Heritage*, Vol. 3, No. 3, winter 1982–83, pp. 2–9.
Snell, F.J., *Chronicles of Twyford* (London: Eliot Stock, 1892).
Sopocy, Martin, *James Williamson: Studies and Documents of a Pioneer of the Film Narrative* (Cranbury, NJ & London: Associated University Presses, 1998).
Souter, Gerry & Janet, *The American Fire Station* (Osceola, WIS: MBI Pub. Co., 1998); *Fire Stations* (New York: Barnes & Noble, 2000).
South Yorkshire Fire Service Historical Society, *Fire and Water: A Collection of Victorian Poems and Advertisements taken from magazines 'Fire and Water' and 'The Fireman' ... from 1880 to 1905* (Sheffield: South Yorkshire Fire Service Historical Society/Hilltop Press, n.d.).
Spehr, Paul C., *The Movies Begin: Making Movies in New Jersey 1887–1920* (Newark, NJ: The Newark Museum, 1977).
Spender, Stephen (writer) & Hinde, John (photographer), *Citizens in War and After* (London: G.G. Harrop & Co., 1945).
St Aubyn, Fiona, *A Portrait of Georgian London based on Rudolph Ackermann's 'The Microcosm of London', Published 1808–1810; Architectural Draughtsman, Augustus Pugin; Caricaturist Thomas Rowlandson* (Churt, Surrey: David Leader, 1985).
Stansky, Peter & Abrahams, William, *London's Burning: Life, Death and Art in the Second World War* (London: Constable, 1994). Includes a chapter on Humphrey Jennings and *Fires were Started*.
Stein, Geoffrey N., *Where's the Fire? American Firefighters in Picture Postcards, circa 1910* (Vestal, NY: Vestal Press, 1991).
Stella, Robert F., *A Pictorial Report on Firefighting* (Boston, MA: Quinlan Press, c. 1987).
Stevenson, John & others, *Yoshitoshi's 'One Hundred Aspects of the Moon'* (Redmond, WA: San Francisco Graphic Society, 1992).
Stoddard, Maynard Good, 'The art of the firefighter,' *The Saturday Evening Post*, Vol. 274, No. 2, March-April 2002, pp. 54–59.
Sumiyoshi, Yamamoto, 'Fire-fighting and disaster prevention in Edo' *Journal of Japanese Trade and Industry*, November/December 2000, published on the Internet: www.jef.or.jp/en/jti/200011_011.html.
Suson, Gary, *Requiem: Images of Ground Zero* (New York: Barnes & Noble, 2002).
Tamarin, Alfred, *Fire Fighting in America* (New York: Macmillan, 1971).
Taylor, Murray A., *Jumping Fire: A Smokejumper's Memoir of Fighting Wildfire* (San Diego, CA: Harcourt Trade Publishers, 2000).
'Terror at the Towers,' *Firehouse*, April 1993.
Tester, Sylvia, *My Friend, The Fireman: A Show and Tell Book* (Elgin, IL: David C. Cook Publishing, 1968).
Thoele, Michel, *Fire Line: The Summer Battles of the West* (Golden, CO: Fulcrum Publishing, 1995).
Tibballs, Geoff, *London's Burning: Behind the Scenes with Britain's Favourite Firefighters* (London: Boxtree/LWT, 1993).
– *London's Burning: Behind the Blaze* (London: Boxtree/LWT, 1995).
Vince, John, *Fire-Marks*, (Aylesbury, Bucks: Shire Publications Ltd., 1973). Shire Album 2.
Wagner, Rob, *Firefighting* (New York: Metro Books, c. 1999).
– *Firehouses* (New York: Metro Books, 2001).
Wallington, Neil, *Fireman – A Personal Account* (Newton Abbot: David & Charles, 1979).
– *Firemen at War: The Work of London's Fire-fighters in the Second World War* (Newton Abbot: David & Charles, 1981).
– *'999': The Accident and Crash Rescue Work of the Fire Service* (Newton Abbot & London: David & Charles, 1987).
– *Images of Fire: 150 Years of Fire-Fighting* (Newton Abbot & London: David & Charles, 1989).
– *Firefighter! The Drama and Humour of a Dangerous Profession* (Exeter: Firestorm, c. 1992).
– *Firefighting: A Pictorial History* (Bristol: Parragon, 1997).
– *Great Fires of London: Images of the London Fire Brigade at Work since 1833* (Stroud, Gloucestershire: Sutton Publishing, 2001).
– & Holloway, Sally, *Fire and Rescue: Into Action with Brigades throughout Britain* (Sparkford: Stephens, 1994).
Wambaugh, Joseph, *The Fire Lover: A True Story* (New York: William Morrow, 2002).
Wanstall, Thomas K., *Fire! At War with the Red Devil* (New York: Bonanza Books, 1989).
Ward, Majorie & Edward, *Danger is their Business* (London: Cassell & Co., 1955).

Ward-Jackson, Philip, 'Sermon Lane ... "Blitz", The National Firefighters' Memorial,' in *Public Sculpture of the City of London* (Liverpool: Liverpool University Press, 2003), pp. 392–94.

Warner, Malcolm, '*Millais, The Rescue*,' in *The Pre-Raphaelites*, ed. by Leslie Parris (London: Tate Gallery Publications, 1984), pp. 131–33.

Wassey, Michael, *Ordeal by Fire* (London: Secker & Warburg, 1941).

Webster, Frank V., *The Young Firemen of Lakeville; or, Herbert Dare's Pluck* (New York: Cupples & Leon Company, 1909).

Wecter, Dixon, *The Hero in America: A Chronicle of Hero-Worship* (New York: C. Scribner's Sons, 1941).

Weight, Angela, *Leonard Rosoman: A War Retrospective* 1939–1945, (London: Imperial War Museum/Edinburgh: The Fine Art Society, 1989).

Weisberg, Gabriel P., *The Realist Tradition: French Painting and Drawing*, 1830–1900 (Cleveland, OH: Cleveland Museum of Art, 1980).

While, John Henry, *Fifty Years of Firc Fighting in London* (London: Hutchinson & Co., 1931).

While, John Henry, *Fire! Fire! A Story of Fire Fighting in Peace and War* (London: Frederick Muller Ltd., 1944).

Wilde, Sally, *Life under the Bells: A History of the Metropolitan Fire Brigade, Melbourne* 1891–1991 (Melbourne: Longman Cheshire, 1991).

Wilke, Jens-Peter (ed.), *150 Jahre Berliner Feuerwehr*, 1851–2001 (Berlin: FKF Media, 2001).

Williams Jr, Herman, with Hall, James, *Firefighter* (Carlsbad, CA: Mountain Movers Press/Hay House Inc., 2002).

Winston, Brian, '*Fires were Started*' (London: BFI Publishing 1999).

Wright, Anthony Allan, *Burning Issues: A Fire Officer's Memoirs of Belfast's Firefighters Baptism of Terrorism during the Troubled Years* 1968–1988 (Donaghadee: Rosepark Publishing, 1999).

Wright, Brian, *Firefighting Equipment* (Princes Risborough, Aylesbury, Buckinghamshire: Shire Publications Ltd., 1989).

– *Firemen's Uniforms* (Princes Risborough, Aylesbury, Buckinghamshire: Shire Publications Ltd., 1991).

– Henham, Brian & Sharp, Brian, *The British Fire Mark*, 1680–1879 (Cambridge: Woodhead-Faulkner, 1982).

Yamamoto, Tadayoshi, *History of Firefighting in Japan* (Tokyo: Fukuinkan-Shoten, 1981).

Yanagi, Munemichi, *Matoi: Symbols of Firemen in the Edo Period* (Tokyo: Unsodo, 1987).

Yokley, Richard C., *TV Firefighters* (San Diego, CA: Black Forest Press, 2003).

Young, Charles Frederick T., *Fires, Fire Engines and Fire Brigades: with a History of Manual and Steam Fire Engines, etc.* (London: 1866).

Zurier, Rebecca (writer) & A. Pierce Bounds (photographer), *The American Firehouse* (New York: Abbeville Press, 1982). Reprinted as *The Firehouse: An Architectural and Social History* (New York: Artabras, 1991).

# Index

**Page numbers in bold italics refer to illustrations**